Podcasting

Podcasting

4th Edition

by Tee Morris and Chuck Tomasi
Foreword by Dr. Pamela Gay

A Wiley Brand

Podcasting For Dummies®, 4th Edition

Published by: **John Wiley & Sons, Inc.,** 111 River Street, Hoboken, NJ 07030-5774, www.wiley.com

Copyright © 2021 by John Wiley & Sons, Inc., Hoboken, New Jersey

Published simultaneously in Canada

For general information on our other products and services, please contact our Customer Care Department within the U.S. at 877-762-2974, outside the U.S. at 317-572-3993, or fax 317-572-4002. For technical support, please visit https://hub.wiley.com/community/support/dummies.

Wiley publishes in a variety of print and electronic formats and by print-on-demand. Some material included with standard print versions of this book may not be included in e-books or in print-on-demand. If this book refers to media such as a CD or DVD that is not included in the version you purchased, you may download this material at http://booksupport.wiley.com. For more information about Wiley products, visit www.wiley.com.

Library of Congress Control Number: 2020944883

ISBN 978-1-119-71181-0 (pbk); ISBN 978-1-119-71182-7 (ePDF); ISBN 978-1-119-71183-4 (epub)

Manufactured in the United States of America

SKY10021685_101420

Contents at a Glance

Table of Contents

Foreword

Welcome to the fourth edition of *Podcasting For Dummies*. This book can be your stepping off point to being part of a vibrant community of content creators who are creating every possible kind of content (often from the quiet of the inside of a closet). While anyone can create a podcast, creating a successful podcast requires showing up week after week and working to constantly improve the quality as you build your audience. No show starts perfect, but this book instructs beginners on how to get a good start while also helping old dogs like me learn new tricks.

I still remember exactly where I was when I saw a real-life copy of the first edition of *Podcasting For Dummies*. The year was 2007. Most people still looked at me funny when I said, "I'm a podcaster," but Tee Morris had recognized me by my voice and told me to come check out a book. Back then, I think that's how all of us original podcasters knew each other: We were voices that we carried around in each other's pockets as we worked to teach each other new content (Mur Lafferty's *I Should Be Writing* still leaves me feeling guilty that I'm not writing!). We also emailed and Skyped, working together to define new techniques and to build up our entire community. Opening that first edition book, I felt a thrill to see my show's logo — *Slacker Astronomy*'s logo — listed side by side with the shows I loved most and embedded in a book that could explain to someone how to create a podcast in the exact way I would create a podcast. (Back then I was a GarageBand + LibSyn + WordPress kind of producer.)

Back in 2007, podcasting was just three years old, but already it felt like everything had changed, as we went from our grassroots origins to mainstream success with the introduction of iTune's Podcast Directory. Already, Tee and Evo were working on a new edition, this time bringing in Chuck Tomasi for added insight. Like me, Chuck was one of the early podcasters, and with his *Technorama* and *Chuck Chat* shows, he interviewed a lot of voices in podcasting, and helped network us together around our craft. Across the years, as podcasting has evolved, this book has evolved, too.

Today, most people know what a podcast is, but creating a stand-out show has actually gotten harder rather than easier. That shouldn't worry you, though. You have this book, and it will introduce you to what's needed to get started. Inside you'll find rock-solid examples of what a good show can look like. Want a science podcast? Got it (That *Astronomy Cast* show? That's me and Fraser Cain!) Want a good tech show? Got those, too. Need insight on production, hosting, promotion? All of it is laid out here. I'm not going to lie. I'll be reading this book to see how I can improve, because 15 years of podcasting later, there are new tools and new best-practices as Spotify, our digital assistants (Hey, Alexa!), and so many other platforms add podcasts to their lineups. There is so much to learn. . . so start reading!

And welcome to the podcasting community. We're so glad you're here.

—Dr. Pamela Gay, Astronomer and cohost of *Astronomy Cast*

Introduction

Maybe you've been casually surfing the Interwebz (yes, that is a thing!) or perusing your newspaper when the word *podcasting* has popped up. Steadily, like a building wave that would make champion surfers salivate with delight, the term has popped up again and again — and your curiosity continues to pique as the word *podcasting* echoes in your ears and remains in the back of your mind as a riddle wrapped in an enigma, smothered in secret sauce.

Well, ponder no more. *Podcasting For Dummies,* 4th Edition provides the answer to the question — What is podcasting? This book takes you through the always-evolving technological movement encompassing the Internet, digital communications, education, and entertainment. By the time you reach the end of this book, the basics will be in place to get you, your voice, and your message heard around the world — and you can even have a bit of fun along the way.

About This Book

Back in 2005 . . .

"So what are you up to, Tee?"

"I'm currently making a podcast of my first novel, a swashbuckling tale that carries our heroes . . ."

"Uh . . . what is a podcast?"

Just the word *podcasting* carries an air of geekiness about it — and behold, the habitual technophobes suddenly clasp their hands to their ears and run away screaming in horror lest they confront yet another nerdy technology that has gone mainstream. Too bad. It has been a long road for podcasting, and it has taken a decade-and-a-half for the platform to go from something nerds are doing in the basement of the science building to a punch line about people who cannot cut it in mainstream media to plot devices in Jordan Peele's reimagining of *The Twilight Zone.* What an evolution!

The funny thing about podcasting — outside of comedy podcasts, of course — is that it isn't that hard to do. When you peel back the covers and fancy-schmancy tech-talk, it's a pretty simple process to make your own podcast. You just need someone pointing the way and illuminating your path.

This is why we're here: to be that candle in the dark, helping you navigate a world where anyone can do anything, provided they have the tools, the drive, and the passion. You don't need to be a techno-wizard or a super-geek. You need no wad of tape holding your glasses together, and your shirt tail need not stick out from your fly. Anyone can do what we show you in this book. Anyone can take a thought or an opinion, make an audio file expressing that opinion, and distribute this idea worldwide. Anyone can capture the attention of a few hundred — or a few thousand — people around the world through MP3 players hiding in computers, loaded on smartphones, strapped around biceps, jouncing in pockets, or hooked up to car stereos.

Anyone can podcast, and anyone can listen.

Podcasting, from recording to online hosting, can be done on a variety of budgets, ranging from frugal to Fortune 500. You can podcast about literally anything — including podcasting for its own sake. As blogging gave the anonymous, the famous, the almost-famous, and the used-to-be famous a voice in politics, religion, and everyday life, podcasting adds volume and tone to that voice.

Podcasting is many things to many people; but at its most basic, it's a surprisingly simple and powerful technology. What it means boils down to a single person: you. This is a platform delivering your message around the world, connecting the Global Village in ways that the creators of the Internet, RSS, and MP3 compression would probably never have dreamed. It is the unique and the hard-to-find content that can't find a place on commercial, college, or public access radio.

You're about to embark on an exciting adventure into digital media distribution, and here you will find out that podcasting is all these things and so much more.

How to Use This Book

Podcasting For Dummies, 4th Edition should be these things to all who pick up and read it (whether straight through or by jumping around in the chapters):

>> A user-friendly guide in how to listen to, produce, and distribute podcasts

>> A terrific reference for choosing the right hardware and software to put together a sharp-sounding podcast

>> The starting point for the person who knows nothing about audio editing, recording, creating RSS feeds, hosting blogs, or how to turn a computer into a recording studio

>> A handy go-to think tank for any beginning podcaster who's hungry for new ideas on what goes into a good podcast and fresh points of view

>> A really fun read

There will be plenty of answers in these pages, and if you find our answers too elementary, we give you plenty of points of reference to research. We don't claim to have all the solutions, quick fixes, and resolutions to all possible podcasting queries, but we do present to you the basic building blocks and first steps for beginning a podcast. As with any *For Dummies* book, our responsibility is to give you the foundation on which to build. That's what we've done our level best to accomplish: Bestow upon you the enchanted stuff that makes a podcast happen.

This book was written as a linear path from the conceptualization stages to the final publication of your work. However, not everyone needs to read the book from page one. If you've already gotten your feet wet with the various aspects of podcasting, jump around from section to section and read the parts that you need. We provide plenty of guides back to other relevant chapters for when the going gets murky.

Conventions Used in This Book

When you go through this book, you're going to see a few ⌘ symbols, the occasional ⇨, and even a few things typed in a completely different style. There's a method to this madness, and those methods are conventions found throughout this book.

When we refer to keyboard shortcuts for Macintosh or Windows, we designate them with (Mac) or (Windows). For Mac shortcuts, we use the Command key and the corresponding letter. For Windows shortcuts, we use the abbreviation for the Control key (Ctrl) and the corresponding letter. So the shortcut for Select All looks like this: Command+A (Mac) / Ctrl+A (Windows).

If keyboard shortcuts aren't your thing and you want to know where the commands reside on menus, we use a command arrow (⇨) to help guide you through menus and submenus. So, the command for Select All in the application's menu is Edit⇨ Select All. You first select the Edit menu and then Select All.

When we offer URLs (web addresses) of various podcasts, resources, and audio equipment vendors, or when we have you creating RSS feeds for podcast clients, such as Apple Podcasts, Overcast, or Spotify, we use `this particular typeface`.

Bold Assumptions

We assume that you have a computer, a lot of curiosity, and a desire to podcast. We couldn't care less about whether you're using a Mac, a PC, Linux, Unix, or two Dixie cups connected with string. (Okay, maybe the two Dixie cups connected with string would be a challenge; a computer is essential.) In podcasting, the operating system just makes the computer go. We're here to provide you tools for creating a podcast, regardless of what OS you're running.

If you know nothing about audio production, this book can also serve as a fine primer in how to record, edit, and produce audio on your computer, as well as accessorize your computer with mixing boards, professional-grade microphones, and audio-engineering software that will give you a basic look at this creative field. You can hang on to this book as a handy reference, geared for audio *in podcasting*. Again, our book is a starting point, and (ahem) a trusted starting point at that.

With everything that goes into podcasting, there are some things this book is not now, nor will ever be, about. Here's the short list:

>> We're not out to make you into an übergeek in RSS or XML (but we give you all you need to make things work — even get you Spotify-ready).

>> We figure that if you get hold of Audacity, GarageBand, or Audition, you can take it from there (but we give you overviews of those programs and a few basic editing examples).

>> We're not out to teach you how to get rich quick through this platform. While we teach you how to produce a podcast, there is no magic formula in making the next *Serial* or *Slow Burn*.

If you are looking for a terrific start to the podcasting experience, then — in the words of the last knight guarding the Holy Grail in *Indiana Jones and the Last Crusade* — "You have chosen wisely."

How This Book Is Organized

The following sections give you a quick overview of what this book has to offer. And yeah, we're going to keep the overview brief because we figure you're eager to get started. But the fact that you're reading this passage also tells us you don't want to miss a detail, so here's a quick bird's-eye view of what we do in *Podcasting For Dummies*.

Part 1: Podcasting on a Worldwide Frequency

Part 1 goes into the bare-bones basics of how a podcast happens, how to get podcasts to your listening device of choice, and how to host a podcast yourself — ending up with a few places online that offer podcast feeds you can visit to sample the experience and (later on) to let the world know "Hey, I've got a podcast, too!"

Part 1 also helps you pick out the best hardware and software you need to start podcasting.

Part 2: The Hills Are Alive with the Sound of Podcasting

Consider this part of the book *Inside the Actor's Studio,* part DIY Network, and part WKRP (with your host, Dr. Johnny Fever . . . *boooouuugaaar!!!*). This is where we offer some techniques the pros use in broadcasting. Podcasting may be the grass-roots movement of homespun telecommunications, but that doesn't mean it has to sound that way (unless, of course, you *want* it to sound that way). From pre-show prep to setting your volume levels to the basics of audio editing, this is the part that polishes your podcast.

Part 3: So You've Got This Great Recording of Your Voice: Now What?

The audio file you've just created is now silently staring at you from your monitor (unless you're listening to it on your computer's music player, in which case it's just defiantly talking back at you!), and you haven't a clue what your next step is. We cover the last-minute details and then walk you through the process of getting your podcast online, finding the right web-hosting packages for podcasts, and putting together show notes that give listeners a glance at what your latest episode is all about.

Part 4: Start Spreadin' the News about Your Podcast

You have your podcast recorded, edited, and online, but now you need to let people know you have this great podcast just waiting for them — and that's what we explore in Part 4. With the power of publicity — from free-of-charge word-of-mouth (arguably the most effective) to investment in social media, you have a wide array of options to choose from when you're ready to announce your presence to the podcasting community.

Part 5: Pod-sibilities to Consider for Your Show

The question of why one should podcast is as important as how to podcast. We cover some basic rationales that many folks have for sitting behind a microphone, pouring heart, soul, and pocket change into their craft each and every day, week, or month. We then look ahead to streaming on Twitch or YouTube as a possible interactive addition to your podcast's workflow.

Part 6: The Part of Tens

Perhaps the toughest chapters to write were these: the *For Dummies* trademark Part of Tens chapters. So don't skip them because we'll be über-miffed if you fail to appreciate how hard we busted our humps to get these chapters done!

Right — so what do we give you in our Part of Tens? Along with giving you ten reasons why podcasting is something important to do, we also offer suggestions for what's out there, how they sound, and how you can benefit from them. Finally, we close with words of wisdom from some of the podcasters who have been podcasting the longest. Read . . . and then you decide.

Authors' note: Keep in mind, the podcasts we cover here are currently active at the time of writing this edition. Some of these podcasts may thrive and continue producing killer content. Some may disappear like ill-fated voyagers tempting the Bermuda Triangle. Remember this as you might find a podcast in these pages that's no longer podcasting.

Icons Used in This Book

So you're trekking through the book, making some real progress with developing your podcast, when suddenly these little icons leap out, grab you by the throat, and wrestle you to the ground. (Who would have thought podcasting was so action-packed, like an MCU movie, huh?) What do all these little drawings mean?

Glad you asked.

When we're in the middle of a discussion and suddenly we have one of those "Say, that reminds me . . ." moments, we give you one of these tips. They're handy little extras that are good to know and might even make your podcast sound a little tighter than average.

If the moment is more than a handy little nugget of information and closer to a "Seriously, you can't forget this part!" factoid, we mark it with a Remember icon. You're going to want to play close attention to these puppies.

Sometimes we interrupt our train of thought with a "Time out, Sparky . . ." moment — and this is where we ask for your completely undivided attention. The Warnings are exactly that: flashing lights, ah-ooga horns, dire portents, or your local DM saying "Roll for Initiative." They're reminders not to try this at home because you'll definitely regret it.

These icons illuminate the "So how does this widget really work . . .?" moments you may have as you read this book. The Technical Stuff icons give you a deeper understanding of what the wizard is doing behind the curtain, making you all the more apt as a podcaster. But if you want to skip the nitty gritty details, that's perfectly fine, too.

Beyond the Book

In addition to what you're reading right now, this product also comes with a free, online Cheat Sheet that offers up quick reference points you'll want to have on hand when you're working on your first podcast, or your first production on location. You need a quick answer at a glance? The Cheat Sheet should jog a memory or two from this book. To get this Cheat Sheet, simply go to www.dummies.com and search for **Podcasting For Dummies Cheat Sheet** in the Search box.

This book also comes with a companion *podcast*. Go to http://www.podcasting fordummies.com and follow the instructions to get free weekly audio commentary from Tee Morris and Chuck Tomasi about concepts in this book explored in greater detail, from the difference between good and bad edits, when too much reverb is too much, and the variety of methods you can use to record a podcast.

Where to Go from Here

At this point, many *For Dummies* authors say something snappy, clever, or even a bit snarky. We save our best tongue-in-cheek material for the pages inside, so here's a more serious approach. . ..

We suggest heading to where you're planning to record your podcast, or just plant yourself in front of a computer, and start with Chapter 1 where you're given a few links to check out, some suggestions on applications for downloading podcasts, and directories to look up where you can find Tee's and Chuck's (many) podcasts, and other podcasts that can educate, inspire, and enlighten your ears with original content.

Where do we go from here? Up and out, my friends. Up and out . . .

1
Podcasting on a Worldwide Frequency

Understand the fundamentals of creating, uploading, and distributing your podcast for others to enjoy.

Explore various podcast directories to find content that interests you and compare and contrast how others produce their shows.

Find the right hardware, software, and accessories to fit your budget.

Take your podcast recording on the road with mobile devices and processes.

Chapter **1**

Getting the Scoop on Podcasting

Sometimes the invention that makes the biggest impact on our daily lives isn't an invention at all, but the convergence of existing technologies, processes, and ideas. Podcasting may be the perfect example of that principle — and it's changing the relationship people have with their radios, music collections, books, education, and more.

The podcasting movement is actually a spin-off of another communications boom: *blogs*. Blogs sprang up right and left in the early 2000s, providing nonprogrammers and designers a clean, elegant interface that left many on the technology side wondering why they hadn't thought of it sooner. Everyday people could chronicle their lives, hopes, dreams, and fears and show them to anyone who cared to read. And oddly enough, people did care to read — and still do.

Then in 2003, former MTV VeeJay Adam Curry started collaborating with programmer Dave Winer about improving RSS (which stands for *Really Simple Syndication*) that not only allowed you to share text and images, but media attachments which included compressed audio and video files. Soon after, Curry released his first podcast catching client. Thus launched the media platform of podcasting.

Podcasting combines the instant information exchange of blogging with audio and video files that you can play on a computer or portable media device. When you make your podcast publicly available on the Internet, you are exposing your craft to anyone with a computer or mobile device and a connection capable of streaming data. To put that in perspective, some online sources report the global online population is over 4.5 billion users. In the U.S. alone, more than 275 million people own some kind of mobile device or portable media player and every one of them is capable of playing your content!

This chapter is for the consumers of the content (the audience) and those who make the content (the podcasters) alike. We cover the basic steps to record a podcast and lay out the basics of what you need to do to enjoy a podcast on your media player.

If you're starting to get the idea that podcasting is revolutionary, groundbreaking, and possibly a major component of social upheaval, great. Truth is, some have made their marks in society. Some storytellers have reignited the desire for short stories, anthologies, and storytelling. Other podcasts have shone spotlights on criminal injustices. But not all podcasts are so deep. In fact, many of them are passion projects inviting you to join in on the experience!

Deciding Whether Podcasting Is for You

Technically speaking, *podcasting* is the distribution of specially encoded multimedia content to subscribed personal computers via the RSS 2.0 protocol. Whew! Allow us to translate that into common-speak:

Podcasting allows you to listen to what you want, where you want, when you want.

Podcasting turns the tables on broadcast schedules, allowing the listener to choose not only what to listen to, but also when — often referred to as *time shifted media*. And because podcasts are transferred via the Internet, the power to create an audio program isn't limited to those with access to a radio transmitter.

The simplest reason to podcast is that it's just plain fun! We've been podcasting since the beginning, and we're still having a blast, continuing to get out messages to our worldwide audiences and challenging ourselves with new tricks and techniques in creating captivating media. So, yeah, for the fun of it? Heck of a good reason.

The following sections cover other reasons podcasting is probably for you.

WHAT'S IN A NAME, WHEN THE NAME IS PODCASTING?

As with most items that make their way into the conventional lexicon of speech, the precise origins and meaning behind *podcasting* are somewhat clouded. Although the domain podcast.com was originally registered back in 2002 (nothing was ever done with it, as far as we know), and Ben Hammersley suggested that and many other terms in February 2004 (www.guardian.co.uk/media/2004/feb/12/broadcasting.digitalmedia), it's generally accepted in the podcast community that the first person to use the term as a reference to the activity we now know as podcasting was Dannie Gregoire on September 15, 2004. Some voices in tech asserted the name held connotations to Apple's popular player of the time, the iPod. Regardless of the intentions, the term was *backronymed* (that is, treated like an acronym and applied to a variety of plausible existing meanings) even with alternative names defiantly offered . . . but to no avail. The term podcasting became part of everyday lexicon.

Since 2004, though, content creators who set up YouTube and Twitch live streams and tell the world "Check out my podcast!" garnered weeping and gnashing of teeth by your authors. We've been what you would call purists, telling people "You can't call it a podcast if you don't have an RSS feed!" However, on finding our own place in streaming, we understood that times change, technology evolves, and even these old dogs can learn new tricks. The term podcast no longer represents the specific technology linked to RSS, but is more about making your content available to a global audience automatically delivered or streaming on-demand. For that reason, we include Chapter 17 to help get you started streaming and seamlessly work it into your podcast's workflow.

You want to deliver media on a regular basis

Sure, you can include audio, video, and PDFs content in your blog if you have one. Many bloggers create special content and insert them as links into the text of their blogposts. Readers then download the files at their leisure. However, this approach requires manual selection of the content blog hosts want readers to download. What sets podcasting apart from blogging is that podcasting automates that process. A listener who subscribes to your podcast is subscribed to all your content, whenever it's available. No need to go back to the site to see what's new! Once you subscribe to a podcast, the content is delivered to you in the same way as when you subscribe to a print magazine. New content, delivered to you. (See what we did there?)

You want to reach beyond the boundaries of broadcast media

In radio, unless it is satellite radio, the number of people who can listen to a show is limited by the power of the transmitter pumping out the signal. Same thing with broadcast television, depending on whether you are using antenna, cable, or satellite dish to receive programming. Podcasting doesn't rely on or utilize signals, transmitters, or receivers — at least not in the classic sense. Podcasts use the Internet as a delivery system, opening up a potential audience that could extend to the entire planet.

No rules exist (yet, anyway) to regulate the creation of podcast content. In fact, neither the FCC nor any other regulatory body for any other government holds jurisdiction over podcasts. If that seems astounding, remember that podcasters are not using the public airwaves to deliver the message.

WARNING

Just because the FCC doesn't have jurisdiction, you're not exempt from the law or — perhaps more important — immune to lawsuits. You're personally responsible for anything you say, do, or condone on your show. Additionally, the rules concerning airplay of licensed music, the distribution of copyrighted material, and the legalities of recording conversations all apply. Pay close attention to the relevant sections in Chapter 5 to avoid some serious consequences. When it comes to the legalities, ignorance is not bliss.

You have something to say

As a general rule, podcasters produce content that likely holds appeal for only a select audience. Podcasts start with an idea, something that you have the desire and knowledge, either real or imaginary, to talk about. Add to that a bit of drive, do-it-yourself-ishness, and an inability to take no for an answer. The point is to say what you want to say, to those who want to hear it.

Podcasts can be about anything and be enjoyed by just about anyone. The topics covered don't have to be earth-shattering or life-changing. They can be about do-it-yourself projects, sound-seeing tours of places you visit, or even your favorite board games. A few rules and guidelines are common in practice, but at times you may find it necessary to bend the rules. (That can be a lot of fun in itself!)

Some of the most popular podcasts are created by everyday people who sit in front of their computers for a few nights a week and just speak their minds, hearts, and souls. Some are focused on niche topics; others are more broad-based.

You want to hear from your listeners

Something that is a real perk with podcasting: accessibility. On average, most audiences have a direct line of contact between themselves and the podcast's host or hosts. Podcast consumers are more likely to provide feedback for what they listen or watch, probably traceable to the personal nature of a podcast. Unlike popular talk shows that follow strict formulaic approaches, podcasts offer their audiences — and the creators behind the production — control, options, and intimacy traditional broadcast media cannot. It is that appeal that attracted major production studios like NPR, ESPN, Disney, HBO, and many others to podcasting. The connection of podcasting with audiences would pave the way for streamers, and now podcasting and streaming are synonymous with one another.

REMEMBER

When you ask for feedback, you're likely to get it — and from unusual places. Because geography doesn't limit the distance your podcast can travel, you may find yourself with listeners in faraway and exotic places. And this feedback isn't always going to be "Wow, great podcast!" Listeners will be honest with you when you invite feedback.

Creating a Podcast

There are two schools of thought when it comes to creating a podcast: the "I need the latest and greatest equipment in order to capture that crisp, clear sound of the broadcasting industry" school of thought, and the "Hey, my laptop has a built-in microphone, and I've got this cool recording software already installed" school of thought. Both are equally valid positions, and there are a lot of secondary schools in-between. The question is how far you're willing to go.

But allow us to dispel a few misconceptions about podcasting right off the bat: You're not reprogramming your operating system, you're not hacking into the Internal Revenue Service's database, and you're not setting up a wireless computer network with tinfoil from a chewing gum wrapper, a shoestring, and your belt, regardless if MacGyver showed you how. Podcasting is not rocket science. In fact, here's a quick rundown of how you podcast:

1. **Record audio or video and convert it to a download-friendly format.**

2. **Write a description of what it is that you just created as a blog post.**

3. **Upload everything to a host server.**

Yes, yes, yes, if podcasting were that simple, then why is this book so thick? Well, we admit that this list does gloss over a few details, but a podcast — in its most streamlined, raw presentation — is that simple. The details of putting together a single episode start in Chapter 2 and wrap up in Chapter 8; then Chapters 9, 10, 11, and 12 walk you through all you need to make the media you create into a podcast.

So, yeah, podcasting is easy, but there's a lot to it.

Looking for the bare necessities

You need a few things before starting your first podcast, many of which you can probably find on your own computer. For these beginning steps, we focus on audio. Here's what you should keep an eye on:

>> **A microphone:** Take a look at your computer. Right now, regardless of whether you have a laptop or desktop model, Windows or Macintosh, your computer probably has a microphone built into it — or a USB port for plugging in a microphone. Yes, even your mobile phone has a microphone, or it wouldn't be much of a phone, now would it? Many earbuds even include a microphone.

Position the microphone in a comfortable spot on your desk or table. If you're using a laptop, it should be somewhere on your desk that allows for best recording results without hunching over the computer like *Young Frankenstein*'s Igor. (That's *EYE*-gor.) Check the laptop's documentation to find out where the built-in microphone is located in the unit's housing. For the mobile phone, hold the device as if you are making a call, the way it was intended. Holding the device any other way can degrade the audio quality. If you are using the microphone earbuds set included with the phone's purchase, you may need to do some experimentation.

TECHNICAL
STUFF

Usually the built-in microphone in a laptop is located close to the edge of the keyboard or near the laptop's speakers. Some models tuck it in at the center point of the monitor's base.

>> **Recording software:** Check out the software that came with your computer. You know, all those extra applications that you filed away, thinking, "I'll check those out sometime." Well, the time has arrived to flip through them. You probably have some sort of audio-recording software loaded on your computer, such as Voice Recorder (PC) or GarageBand (which comes pre-installed with new Macs).

TIP

If you don't already have the appropriate software, here's a fast way to get it: Download the version of Audacity that fits your operating system (at https://audacityteam.org), shown in Figure 1-1. (Oh, yeah . . . it's free.)

>> **An audio interface:** Make sure your computer has the hardware it needs to handle audio recording and the drivers to run the hardware — unless, of course, you have a built-in microphone.

Some desktop computers come with a very elementary audio card built into the motherboard. Before you run out to your local computer vendor and spring for an audio card, check your computer to see whether it can already handle basic voice recording.

For tips on choosing the right mic and audio accessories, be sure to check out Chapter 2. Chapter 3 covers all the software you need.

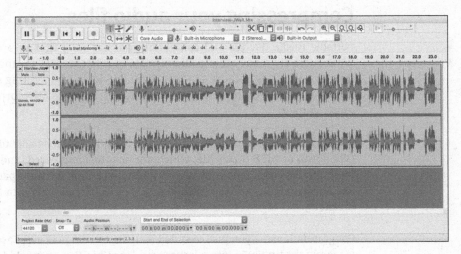

Recording your first podcast

When you have your computer set up and your microphone working, it's time to start recording. Take a deep breath and then follow these steps:

1. **Jot down a few notes on what you want to talk about.**

   Nothing too fancy — just make an outline that includes remarks about who you are and what you want to talk about. Use these bullet points to keep yourself on track.

   All this — checking your computer, jotting down notes, and setting up your recording area — is called *preshow prep,* discussed in depth in Chapter 5 by other podcasters who have their own set ways of getting ready to record.

2. **Click the Record button in your recording software and go for as long as it takes for you to get through your notes.**

TIP

We recommend keeping your first recording to no more than 20 minutes. That may seem like a lot of time, but it will fly by.

3. **Give a nice little sign-off (like "Take care of yourselves! See you next time.") and click the Stop button.**

4. **Choose File ⇨ Save As and give your project a name.**

Now bask in the warmth of creative accomplishment.

Compressing your audio files

Portable media devices and computers can play MP3 files as a default format. While there are many other audio formats in existence, MP3 is the preferred format for podcasting because so many digital devices and operating systems recognize it. If your recording software can output straight to MP3 format, your life is much simpler.

If your software cannot export directly to MP3, it should be able to save to a WAV (Windows) or AIFF (Mac) file, which are raw, uncompressed, and can get rather large. In this case, save your raw file from your first software package and then use Audacity to import the file and export it as MP3. We get into those details in Chapter 3.

Congratulations — you just recorded your first audio podcast! Easy, isn't it? This is merely the first step into a larger world, as Obi-Wan once told Luke.

Uploading your audio to the web

An audio file sitting on your desktop, regardless of how earth-shattering the contents may be, is not a podcast. Nope, not by a long shot. You have to get it up on the Internet and provide a way for listeners to grab that tasty file for later consumption.

If you already have a web server for your blog, company website, or personal website, this process can be as easy as creating a new folder and transferring your newly created audio file to your server.

If that last paragraph left you puzzled and you're wondering what kind of mess you've gotten yourself into . . . relax. We don't leave you hanging out in the wind. Chapter 10 covers everything you need to know about choosing a web host for your podcast media files.

TECHNICAL STUFF

A podcast *media* file can be any sort of media file you like — audio, video, or even an interactive PDF. While our primary attention is on audio, you can use all the tips we give here to handle other types of media.

After you upload your episode, you need to have an RSS (*Really Simple Syndication*) file generated to deliver it. That happens automatically on a blog. The RSS describes where to find the media file you just uploaded. Nearly all software for blogging (called *blog engines*) support RSS, but not all support podcasts. You may need to add a *plug-in*, a downloadable extension that make podcast support a simple affair. This generated RSS file is your *podcast feed*. People who listen to your podcast can subscribe to your show by placing a link to this podcast feed in their podcast application, such as Apple's Podcast, Overcast (shown in Figure 1-2), Stitcher, and Spotify. All these apps are looking for your podcast's RSS, and list it in their *directory*, which is exactly what it sounds like — a digital catalog of podcasts that are available to you.

FIGURE 1-2:
Podcasting can happen in various kinds of media. Whether it is audio (seen here), video, or interactive PDFs, your podcast can deliver content relevant to your audience.

Yes, we know . . . this all sounds really complicated. But we assure you, it's not. Some hosting companies, such as LibSyn (https://libsyn.com), specialize in taking the technological "bite" out of podcasting so that you can focus on creating your best-sounding show. With LibSyn (shown in Figure 1-3), moving your file to a web server is as simple as pushing a few buttons while the creation of the RSS 2.0 podcast feed and even the accompanying web page are automatic.

FIGURE 1-3:
LibSyn handles many of the technical details of podcasting.

If you want to take more control over your website, podcast media files, and their corresponding RSS 2.0 feed, look at Chapters 9 and 10. In those pages, we walk you through some essentials — not only how to upload a file but also how to easily generate your RSS 2.0 file using a variety of tools.

Grabbing listeners

With media files in place and an RSS 2.0 feed ready to be recognized by your podcast app of choice, you're officially a podcaster. Of course, that doesn't mean a lot if you're the only person who knows about your podcast. You need to spread the word to let others know that you exist and that you have something pretty darned important to say.

Creating show notes

Before you pick up a bullhorn, slap a sandwich board over yourself, and start walking down the street (virtually, anyway), you have to make sure you're descriptive enough to captivate those who reach your website. First, you're going to want to describe the contents of your show to casual online passersby in hopes of getting them to listen to what you have to say. That blog post you created to help deliver your media is that place. This is where *show notes* take form and give people a rundown of what you're talking about.

REMEMBER

You can easily glance at a blog and get the gist of a conversation, but an audio file requires active listening to understand, and it's quite difficult to skim. In effect, you're asking people to make an investment of their time in listening to you talk, read a story, or play music. You need some compelling text on a web page to hook them.

Show notes are designed to quickly showcase or highlight the relevant and pertinent contents of the audio file itself. A verbatim transcript of your show isn't always necessary, but we do recommend more than simply saying "a show about my day." Chapter 11 discusses ways to create your show notes and offers tips and tricks to give them some punch.

Getting listed in directories

When you have a final media file and a solid set of show notes, you're ready to take your podcast message to the masses. You can get listed on some directories and podcast-listing sites, such as Spotify, Apple Podcasts, Google Play, Stitcher, and BluBrry (explained later in this chapter). Potential listeners visit literally dozens of websites as they seek out new content, and getting yourself listed on as many as possible can help bring in more new listeners to your program.

WARNING

A huge listener base is a double-edged sword: More demand for your product means more of a demand on you and the resources necessary to keep your podcast up and running. We recommend working on your craft and your skills, as well as getting a good handle on the personal and technological requirements of podcasting, before you embark on a huge marketing campaign. When you're ready, Part 4 has more details about marketing. Part 4 spends a lot of time talking about the various ways you can attract more listeners to your show and ways to respond to the ideas and feedback that your listeners inevitably provide. Many podcasters are surprised at the sheer volume of comments they receive from their listeners — but when you consider how personal podcasting is (compared to traditional forms of media distribution), that's really not surprising at all.

There's an App for That

You have the media file, an RSS feed, and accompanying show notes. You're all set, but ask yourself, "How do podcasts get from the web to my device so I can watch or listen?" To access all this great, new content, you need a *podcatcher,* an application that looks at various RSS feeds, finds the new stuff, and transfers it from the Internet to your computer or mobile device automatically. In this section, we take a look at some of the different podcatching apps available for your listening/viewing needs.

REMEMBER

You may think you need an iPod for all kinds of reasons, but you really don't need one to podcast. Allow us to state that again: You do not need an iPod to listen to or create a podcast. You don't even need an iPhone to listen to or create a podcast, either. The only "I" you need is yourself. As long as you have an MP3 player — be it an application on a Mac, an application on a PC, mobile phone, or any portable device you can unplug and take with you — you possess the capability to listen to podcasts. Depending on the MP3 player, you may even be able to create your podcast on the device as well, but to listen, all you need is a device that can play audio files. This includes your computer.

The following sections represent only a starting point for getting access to podcasts. Any attempt at a comprehensive list would be instantly obsolete. Podcasting continues to grow in popularity, and new podcasting apps are coming out all the time. And remember, you can listen to podcasts on all sorts of devices besides computers — smartphones, tablets, AppleTV, Roku, and more!

The old-timer: Apple Podcasts

With iTunes launching its own podcast directory and a podcast-ready version of its player in June 2005, podcasting went from what the geeks were doing in the basement of the Science Building to the next wave of innovation on the Internet (which was, of course, developed by the geeks in the basement of the Science Building). Plenty of contributing factors helped push podcasting mainstream, but iTunes introducing a push-button subscription method was a huge step forward. As always, such a step into the mainstream market brought some dismay to the original podcasters, finding themselves overshadowed by larger media entities. Now, recognizable giants like NPR, *The New York Times*, ESPN, BBC, and so on dominate Apple's podcast directory (shown in Figure 1-4). What about the indie podcasts — the ones that started it all? Would they be forgotten? Go unnoticed? Languish unsubscribed? Well, at first, it seemed that many of the original ground-breakers that the podcasting community knew and loved (*Comedy4Cast, Evil Genius Chronicles, Coverville, GrammarGirl*) might get lost in the stampede. But not yet, as it turns out.

If you go looking for iTunes on your Mac and cannot find it, it could be that you have upgraded to Catalina (2019) or later. The monolithic iTunes app has been replaced with a separate podcasts app (along with another for music and a third for shows and movies). As for the Windows edition, fear not — as of this writing, you can still get iTunes for Windows (available for download at www.apple.com/itunes). Whether using Apple Podcasts or iTunes on your desktop, the software lends an automatic hand to people new to podcasting, where to find blogs that host podcasts, and which podcast directories list the shows that fit their needs and desires — now they, too, can enjoy a wide range of podcast choices.

FIGURE 1-4:
The Apple
Podcast directory.

Subscribing to podcasts using Apple Podcasts is easy. Just follow these steps:

1. **Find the podcast of your choice.**

    You can do that by

    - Browsing the list podcasts from the Browse option on the left side.

    - Viewing the popular podcasts by using Top Charts below that.

    - Searching by using the Search box in the upper left.

2. **When you find a show that you are interested in, click the image to get to the podcast page and then click the Subscribe button.**

After your podcast finishes downloading, you can find the new episodes by going to the Listen Now section in the left column of Podcasts, or get specific shows or episodes from your library (also on the left).

Podcasting on the go: Stitcher

As smartphones and tablets became more and more prevalent, the notion of having "an app for that" grew in demand. Finding and subscribing to podcasts in the early days was possible on a mobile device, but a bit clunky. The stage had been set for a new kind of interface that was just as easy and elegant to use on your mobile device as it would be on a computer.

Enter Stitcher.

Stitcher Radio (http://www.stitcher.com) with an app available for both iPhone and Android) debuted in 2008 and has quickly become a must have for not only podcasts, but for radio shows coast-to-coast and around the world (see Figure 1-5). On the app or on their computer, members of the Stitcher community "stitch" together on-air feeds from radio stations and podcasts both from professional and amateur studios into personalized stations all readily available on your mobile devices.

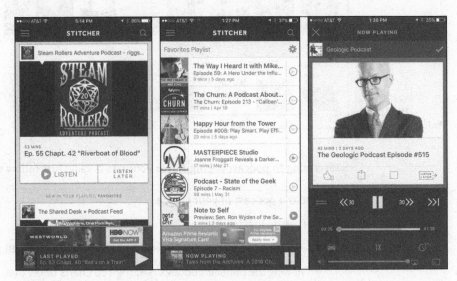

FIGURE 1-5:
Stitcher Radio, bringing podcasts and talk radio to your mobile device in a feed that customizes to your listening habits (left), can organize subscriptions on a Favorites list (middle), and gives you control over the episode playing through your smart device (right).

Once you set up a free user account on Stitcher, finding a podcast for your personal feed is only a few taps away:

1. **Launch Stitcher on your mobile device.**

2. **Sign in with either Facebook, Google, or your Stitcher account.**

3. **Find the podcast of your choice.**

   You can do that by

   - Tapping on the Search icon in the bottom right corner.

   - Swiping up through the offered feeds on your Front Page. (The more you use Stitcher, the more shows matching your interests will appear here.)

4. **When you find something you like, tap the show's image.**

5. **To subscribe, tap the plus sign (+) just below the show's image.**

   The show appears in your favorites list (see Figure 1-6). Click the play icon next to any episode to listen right now.

Stitcher is not only a terrific option for podcast audiences as the service is available on a variety of platforms, including playback through your browser, but Stitcher also interfaces seamlessly with many makes and models of automobiles. Plugging your smartphone into your car's USB port is immediately recognized, and your car's media center immediately picks up where you left off.

Welcome the game-changer: Overcast

When it came to Apple and podcasting, the early days of the Apple Podcasts app were a bit tumultuous. Playback was awkward, interfaces were less than intuitive, and crashing was not uncommon. In 2013, developer Marco Arment introduced a modest app designed for subscribing and listening to podcasts at the XOXO festival. Ironically, he unveiled the app during a talk on competition and the risks and benefits of how crowded the creative field had become. It's ironic as this app, Overcast (`https://overcast.fm`), received rave reviews with *The Verge* and *The Sweet Setup* calling it "the best podcast app for iOS" for many years.

Along with its stability, Overcast offers what podcasting apps of that time lacked: simplicity. Subscribed podcasts were listed alphabetically. Going into a podcast, episodes could be listed by unplayed episodes or all episodes available, as seen in Figure 1-7. The settings for the app and even the podcast itself could be controlled from the app. You can also play episodes back faster or slower than their original speed, and you can pick up from where you leave off.

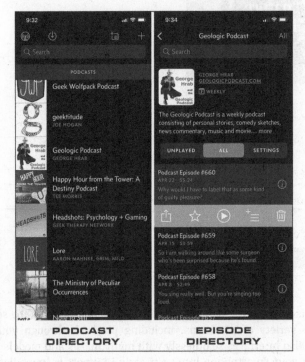

FIGURE 1-7:
Overcast is an app that offers a simple interface in finding your subscribed podcasts (left) and organizing your podcasts.

Once you set up your free account and download the app, finding a podcast for your personal feed is only a few taps away:

1. **Download and launch Overcast on your mobile device.**

2. **Sign in with your account.**

3. **Find the podcast of your choice.**

   You can do that by

- Tapping on the Search Directory field across the top.

- Scanning the various podcasts offered in Overcast's categories.

- Adding a podcast's URL manually by tapping the Add URL option in to the top-right corner.

4. **When you find something you want to know more about, tap the podcast.**

5. **Once you find a podcast you want to listen to, tap the Subscribe button.**

Arment kept things simple in this app's UI (short for *user interface*) and created an elegant interface that makes Overcast a "must-have" for those on the iOS platform, even with Apple Podcasts available everywhere.

Podcasting with the G-man: Google Play Music

Not to be outdone by Apple and the iOS developers, the diversified search engine service Google, the people who bring you Google Drive, Google Docs, and Google Voice (which we discuss later on), offers Google Play Music (at `https://play.google.com/store`) for Android users. It is exactly as it sounds: Google's version of the Apple iTunes (before the breakup), all in one.

Google Play, shown in Figure 1-8, works with the Google Play Music app to bring podcasts to you in the same way as Stitcher and Apple Podcasts. A few extra steps are involved in getting podcasts subscribed and playing on your phone. Once you download Google Play Music and sync it up with your Google account:

1. **Launch Google Play Music on your mobile device and then tap on the Menu option in the top left of the app screen.**

2. **Select Podcasts from the options.**

3. **Find the podcast of your choice.**

   You can do that by

   - Tapping on the Search icon in the top right corner side.

   - Tapping on the All Categories option to reveal the top rated podcasts of categories and subcategories of interest.

   - Swiping through Top Charts, podcasts that are either promoted or popular. This is the main page of Google Play Music's Podcasts directory.

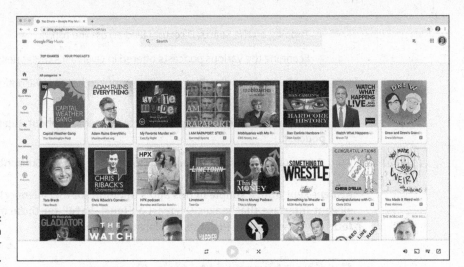

FIGURE 1-8:
Google Play, an
alternative for
Android users.

4. **When you find the podcast you want to listen to, click the Subscribe button.**

Google Play Music offers a series of options for you to complete before subscribing.

- Select Auto-download if you want the latest content to update automatically.

- Opt-in for a push notification when new content downloads.

- Select a Playback Order if you want podcast episodes to go from oldest to newest, or newest to oldest.

5. **Choose your option and tap Subscribe again.**

Google Play Music's UI is targeted mainly towards music, although with podcasts, it gets the job done. Granted, if your loyalties are not to one operating system, Spotify works for both iOS and Android users. Regardless of which app best suits you and your digital lifestyle, it's good to know options are out there, and available for you.

A new 800-pound gorilla: Spotify

For nearly a decade, the king of podcasting app was Apple iTunes, but like all technology, the crown must get passed. When you say Spotify, most people will think of streaming music, but in 2015 it added support for podcasts. It was exclusively set aside for music artists and broadcast media productions, and in 2018, Spotify simplified its application and screening process and now brings podcasting to millions of listeners worldwide (see Figure 1-9). The desktop app requires you to make a download from Spotify's site (`https://spotify.com`). Once installed, log in or create a free account.

1. **Use the Search box in the top left to enter a podcast name or click Browse on the left and select Podcasts.**

2. **When you find a podcast you like, take one of the following actions:**

    -   Click the play icon to listen immediately.

    -   Click the heart icon to save it to your library.

    -   Use the three-dot menu and choose Follow. This is Spotify's way of subscribing to a podcast.

    Now you can quickly find the podcasts you follow by clicking Podcasts on the left under Your Library.

FIGURE 1-9: After years of offering music from all artists of all genres, Spotify now brings podcasts to millions of subscribers.

The Spotify mobile experience is very similar:

1. **After signing in, tap Search on the bottom to search and browse podcasts.**

2. **Tap the Follow button or listen to an episode directly from the list.**

3. **Followed podcasts are available from Your Library at the bottom of the screen and tapping Podcasts (in big letters at the top).**

To listen anywhere, click the down arrow on any episode, and it will be available in your Downloads list under Podcasts.

WARNING

At the time of this writing, Spotify only approves audio podcast feeds for its directories. Any episodes that are offering other forms of media will be rejected. In order to get your feed approved for Spotify, you will need to assure that all your episodes are audio. You can still have posts on your blog featuring embedded videos and special downloads, but you will not be able to use the RSS feed as a delivery method.

Other Podcast Resources

It goes without saying that this book is a snapshot in time, and you will likely want to keep up on the latest news and information of the podcasting world. There are plenty of resources online and in meat-space (real live human beings) where you can ask questions and exchange information. A quick search of LinkedIn, Meetup.com, Facebook, and yes, even Google, can reveal a wealth of resources after you're done reading this book and listening to our podcast.

Chapter **2**

Getting the Gadgets To Produce a Podcast

N ow that you're effectively hip as to what podcasting is and what it's all about, you need to start building your studio. This studio can come in a variety of shapes, sizes, and price tags. What you're thinking about doing — podcasting, in this case — is like any hobby you pursue. If it's something to pass the time, then keep your setup simple. A modest setup with little to no investment is ideal if you want to see whether you like playing with audio. If, on the other hand, you find yourself tapping into a hidden passion or (even better) a hidden talent, you might want to upgrade to the audio gear that's bigger, better, and badder than the basics.

If you suddenly decide you — *yes, you* — have a message you want to share, your next plan of action will be one of two options: picking up a digital recorder, reading the instruction manual, and then downloading some free audio-editing software; or watching the DIY Network for methods of soundproofing the basement and looking at industry-standard equipment that might require some extra homework to master.

This is the beauty of podcasting. In the long run, it doesn't matter whether your podcasting studio is a smartphone with a plug-in microphone or comprised of the latest mixing boards, audio software, and recording equipment. Both approaches to podcasting work and are successfully implemented on a variety of podcasts.

So which one works best for what you have in mind? That's what we take a look at in this chapter, discussing the options, advantages, and disadvantages of each setup.

Finding the Right Mic

It's easy and affordable to make your computer podcast-capable. Your first order of business is to find the right mic. If you already have a microphone built into your laptop (and you don't mind starting out basic), you can just download Audacity (described in Chapter 3) and take advantage of the fact that it's free. That's all there is to it!

At least, for starters, anyway.

However, if your desktop computer didn't come with a mic or you're just not happy podcasting Quasimodo-style, you're going to want to shop around. While microphone shopping, consider the following criteria:

>> **What's your price range?** In many cases, especially with "cheap" mics, you get what you pay for in quality of construction and range of capabilities. You want the most affordable model of microphone that will do the job for you and your podcast.

>> **Do you plan to use the mic primarily in the studio or on location?** A high-end shock-mounted model isn't the best choice for a walk in the park, and a lapel mic might not provide you with the quality of sound expected for an in-studio podcast. Before you purchase your mic, consider your podcasting location needs.

REMEMBER

There are many different types of microphones and microphone connections or *jacks*. Some mics are built to plug in to mixers while others connect directly to your computer's USB port. If you're thinking of starting with an inexpensive mic, you may not be able to plug it in to your upgraded rig later on. Similarly, if you're upgrading to an XLR microphone (described later in this chapter), you may find yourself upgrading more than just your mic.

>> **What do you want this mic to do?** Pick up sound? Yeah, okay, but what kind and in what surroundings? Some mics offer ease of handling for interviews. Some are good at snagging specific outdoor sounds. Others may be better at capturing live music. You need to consider multipurpose mics, guest mics, and on-location recording devices, or any kind of condition your podcast may require.

Even after narrowing your options, there are so many microphones on the market. After a while, the manufacturers, makes, and models all start to give you that kind of brain-freeze you get when you eat ice cream too quickly. The sections that follow give you the lowdown on the mic that's right for your budget and even make a few sound recommendations along the way. Sound recommendations! Heh, Dad humor.

TIP

Say, did you know there's a companion podcast for this book? Yeah, it's *Podcasting For Dummies: The Companion Podcast.* In Season 4, Tee and Chuck talk at length about microphones, including testing several makes and models, comparing and contrasting the sound of each. To find out more about microphones, gadgets, and concepts, listen to *Podcasting For Dummies: The Companion Podcast,* available at http://www.podcastingfordummies.com.

A Beginner's Guide to Mics

The microphone market, over the years, has seen the Universal Serial Bus (USB) microphone go from trendy tech gimmick to state-of-the-art innovation changing the way we record audio. With more than 15 years of development, one consistency has been the setup of these microphones: Plug it in and start recording. That's it. You can go online to any of the computer equipment retailers and type **USB microphones** in their respective search engines. Make sure you include **USB**, because a search for microphones can bring up many more alternatives, including video cameras, high-end mics (which we talk about next), and other devices that may be way out of your budget.

REMEMBER

You don't have to sacrifice your retirement fund to get started with a USB microphone. Prices start under $20 for a simple desktop microphone. The phrase "You get what you pay for" doesn't always apply in podcasting either. We've heard some pretty amazing sounds out of inexpensive microphones.

WARNING

Built-in microphones and many of the cheaper mics are in physical contact with the desk/table in front of you which means they pick up vibrations. Any noise on the surface, such as setting down a glass or drumming fingers, is going to be amplified in the microphone and recorded. Check out the "Accessorize! Accessorize! Accessorize!" section, later in this chapter, where we talk about mic stands to help reduce this issue.

You can find USB microphones online starting as low as $10. For a few dollars more, you can invest in a gaming *headset* like the Logitech G Pro X, pictured in Figure 2-1. The Pro X headset, priced at $130, is a standout among other gaming headsets as its mic uses the same technology found in Blue Microphones (https://www.bluedesigns.com), a pioneer in USB microphone development. This headset provides podcasters with a simple, single-purchase solution offering exceptional audio quality in the headphones coupled with clean, clear pickups only found in Blue mics.

FIGURE 2-1:
The LogitechG Pro X gaming headset is equipped with a microphone running Blue Microphone technology, providing crystal clear audio in both monitoring and recording.

One advantage headsets hold over a desktop microphone is that your mouth is always the same distance away from the actual part of the microphone that picks up sound. If you're an animated podcaster like your authors, this may be useful. On the other hand, headsets can be very sensitive to breathing sounds. If the headset's mic boom is in the wrong position, the slightest breath from your mouth or nose will sound like an EF5 tornado passing by.

REMEMBER

Although purchasing a USB headset is a monetary plus because you're getting both headphones and mic together with one purchase, this isn't all you need to monitor yourself as you record. To self-monitor your recording, you need software that offers you an on-screen mixer, a mixing board, or the option to monitor the incoming audio signal.

Once you plug in your mic or headset into a USB port and set up your audio preferences in your recording software. You're ready for recording.

WARNING

While USB microphones are quick and easy to set up, there's no real good way to connect more than one to a computer at a time. If you're planning to have your best friend as a cohost or have an in-studio interview guest, then consider microphones that connect to a mixing board.

Investing in a high-end mic

So you can go cheap and pick up a microphone for as little as ten dollars, but let's be honest here: You're going to sound like you've been recorded by a ten dollar microphone. Say, however, you really want that sharp, professional sound for your podcast, and the cheaper microphones just aren't cutting it for you. As you shop for an upgrade, you see mics ranging in price from an inexpensive $70 and reaching up to $20,000! (No, you're not seeing a typo involving a couple extra zeroes.) So what defines a microphone? Price? Manufacturer? Look?

What truly defines a microphone is how you sound on it and how it reproduces the sound coming in. Prices vary on microphones based on how they work. Plenty of high-quality microphones out there can pick up nuances and details and still remain in the range of affordability.

WARNING

When you purchase a higher-end microphone, keep in mind that you probably will not receive cables for hookup, a jack that fits into your computer or a stand. That's because you're upgrading to professional equipment. The manufacturer is offering you the flexibility to have different bells and whistles, the whatjamajigs, and extra doodads to make this mic work. For the lowdown on what accessories you need to hook up your new mic to your computer, check out the "Accessorize! Accessorize! Accessorize!" section later in this chapter.

Remember those three questions a few pages back? Question #2 — *Do you plan to use the mic primarily in the studio or on location?* — helps narrow your search even more for the microphone that's the right investment for you. At this level, there are two kinds of mics you will hear people talk about: *dynamic* and *condenser* mics.

Dynamic mics

Dynamic mics are what you see everywhere from speaking engagements to rock concerts. In fact, when someone mentions the word "microphone," the image that comes to mind is probably a dynamic microphone. These mics work like a speaker in reverse. Sound entering a dynamic mic (by speaking directly into it) vibrates a *diaphragm* (a small plate) attached to a coil. This is located within proximity of a magnet, and the vibrations that this Wile E. Coyote setup makes create a small electric current. When this signal runs through a preamp or mixer, the original sound is re-created. This system sounds complicated (and if you've ever looked inside of a microphone, it is), but the internal makeup of dynamic microphones is such that they can take a lot of incoming signal and still produce audio clearly. They're also rugged in build so they can be manhandled, making dynamic mics exceptional for outdoor recording.

If you're working in-studio, consider working with Shure's SM7B (pictured in Figure 2-2). This mic offers up a clean, rich sound, and picks up incredible details for a dynamic microphone. For microphones you may want to use out in the field, the Røde Reporter (http://www.rode.com/microphones/reporter) is specifically designed for handheld interviews, and delivers broadcast-quality results within any environment.

FIGURE 2-2: The Shure SM7B, a standard in the professional audio industry, captures both high and low audio ranges, giving your podcast a polished edge.

Condenser mics

When podcasting happens in studio and you're looking for the subtleties and nuances of the human voice in your recording (the more detail you get, the better!), you may want to shop for studio *condenser* microphones. The anatomy of a condenser microphone is similar to a dynamic one. In the condenser, a diaphragm is suspended in front of a stationary plate that conducts electricity. As a signal enters the microphone, the air between the diaphragm and the plate is displaced, creating a fluctuating electrical charge. Once given a bit more power (*phantom power,* which is explained later in this chapter), the movement becomes an electrical representation of the incoming audio signal.

This setup sounds very delicate, doesn't it? Guess what — it is! This is why condenser mics are transported in padded cases, and not really built for handheld use. These microphones are best used for in-studio recordings (as shown in Figure 2-3) versus on-location podcasts. If manhandled or jostled around, plates can be knocked out of alignment or damaged, causing problems in the pick-ups.

FIGURE 2-3:
The MXL 990 captures professional quality sound in home studio settings for a reasonable investment.

The advantage to this delicate setup is that condenser mics are far more sensitive to sound, and they pick up a wider spectrum of audio. These microphones are so sensitive to noise around them that some come with *shock mounts* — spring-loaded frames that "suspend" the mic when attached to a microphone stand, providing better stability while reducing any noise or vibration from your microphone stand. Think of a shock mount as a shock absorber for your mic.

REMEMBER

Although dynamic microphones are marketed more for on-location recording and condenser mics are considered best for in-studio recording, these aren't hard-and-fast rules for what mics should be used where. Sometimes, podcasters use condensers in outdoor settings, and some podcasters prefer the sound of dynamic mics in studio over condensers. When picking a microphone, you want a mic that not only suits your needs but also makes you sound good. *Really* good.

TIP

For multi-mic podcasts in the same location, consider dynamic microphones. If you plan to have a studio with a few friends over as guests, the sensitivity of *several* condenser microphones can be a liability and will degrade the overall sound of your podcast. The sound from one person's voice will be picked up not only on their mic, but also on the other mics as a "distant" sound. Overall it will sound like everyone is talking in to a single microphone across the room like one of those bad conference room calls with the mic in the center of the table. Very distracting to the listener.

Tee is a big fan of MXL microphones (http://www.mxlmics.com), mentioned earlier in this chapter. His first in-studio microphone was an MXL 990, a model he still podcasts with to this day. MXL microphones, along with being affordable, reliable, and popular, often come bundled with *mixer boards* (discussed later in this chapter). Online vendors like BSW (http://www.bswusa.com) are offering podcast bundles featuring another MXL microphone, the BCD-1. With each bundle offering new accessories, cost and features go up, but the high quality sound that the MXL line captures remains the same.

USB Studio condenser mics

Microphone vendors are noting the popularity of podcasting, and now *USB Studio Condenser* microphones are becoming more and more prevalent. Leading this movement is the previously mentioned Blue Microphones (https://www.bluedesigns.com). After the success of the Snowball iCE ($70) in the early years of podcasting, Blue went on to develop the Yeti ($130), both reliable examples of what is now available for podcasters. Blue also offers the Yeti X ($170), an impressive audio solution for podcasters, offering mixing, gain, and metering all in one simple interface built into the microphone. The Yeti X, shown in Figure 2-4, also provides a variety of pickup patterns, making the microphone versatile in recording settings and situations.

With these microphone models, podcasters can now record studio quality audio without an additional audio card, a mixer, or any of the go-betweens once considered essential for connecting audio gear to a computer. With USB condenser microphones, the audio signal now has a direct connection to the computer. It makes your podcast production extremely portable.

THE PHANTOM (POWER) OF THE PODCAST

You might notice that condenser mics (such as the MXL microphones described in this chapter) are *phantom powered,* their connection to the mixer provides them with all the power they need. Because phantom power comes from the mixing board (in the studio) through the cable connecting to the mic, it supplies a constant boost. Portable condenser mics use batteries to get their kick, but when they start to die, so does your recording quality. (A word to the wise: Always keep fresh batteries on hand and check your mic's battery level before you record.) Similar considerations should be taken when you are using a portable recorder as a *pre-amp,* covered in Chapter 4.

FIGURE 2-4:
The *Yeti X* yields professional level audio and features, all in one sleek, elegant design. Within minutes of its unboxing, you are connected to your computer and ready to record.

BUNDLING UP

While many microphones offer a single-purchase solution for podcasters, the microphone itself is not the only thing you need. Maybe you need a stand? What about a longer cable to accommodate your home studio? You could go on and put together a list of accessories that you need for your centerpiece . . .

. . . or follow the lead taken by a few audio vendors and look at the *bundles* they offer.

Blue Microphones offers the *Yeticaster* (https://www.bluedesigns.com/products/yeticaster/) as the ultimate solution when you're ready to upgrade to a professional line of USB mics. The Yeticaster features the Yeti microphone, a Radius III shock mount designed specifically for Yeti models, and the Compass microphone boom arm offering hidden-channel cable management. MXL has put together the Overstream bundle (http://www.mxlmics.com/microphones/podcasting/overstreambundles/), available in either *Blaze* or *Blizzard* options. The bundle includes the MXL 990, a shock mount with integrated pop filter, a Mic Mate Pro, and a boom arm ready for mounting on any table or desk. With the Mic Mate Pro, the 990 can be either a USB mic or an XLR, depending on your needs.

(continued)

(continued)

Take a look at the different bundles that vendors are now putting together for microphones, headphones, and mixers. You may just find that one purchase will be all you need to get started on your podcast.

And if you want to know more about the accessories in the bundles featured here, refer to the "Accessorize! Accessorize! Accessorize!" section, later in this chapter.

REMEMBER

High end is relative. Professional recording studios can spend hundreds of dollars on a single microphone. Fortunately, the MP3 format isn't so finicky that modest-priced microphones can sometimes stick the landing. For those of you wanting to invest in your studios, we're defining *high-end* microphones in this book as microphone makes and models ranging from $150 and up.

Podcasts Well with Others: The Mixing Board

If you're looking to have more people in studio or you want to entertain guests on your podcast, you need more than just a really good microphone. (Otherwise there will be the passing back and forth of a mic, and that'll get distracting after a while.) You also need inputs for more microphones and full control over these multiple microphones.

Along with good microphones (and probably *microphone stands*, discussed later in the "Accessorize! Accessorize! Accessorize!" section), you will want to invest in a *mixing board*. What a mixing board (or *mixer*) does for your podcast is open up the recording options, such as multiple hosts or guests, recording acoustic instruments, and balancing sound to emphasize one voice over another or balance both seamlessly.

You see mixers at rock concerts and in "behind the scenes" documentaries for the film and recording industry, appearing in all shapes and sizes. For example, the Behringer Q1202USB mixer, priced around $150, can connect four microphones and several additional stereo inputs and feed these audio sources directly in to the USB port on your computer.

TIP

Before you take your mixer out of its packaging, make sure you have room for it somewhere on the desk or general area where you intend to podcast. Something you'll want to consider is how close you want to be to the mixer. An ideal reach for your mixer is a short one. Whether it's you recording yourself or balancing the levels of those around you, clear off a section within a relaxed reach. You'll be happier because of it.

The anatomy of a mixing board

The easiest way to look at a mixing board is as if you're partitioning your computer into different recording studios. But instead of calling them *studios*, these partitions are called *tracks*.

A mixing board provides mono tracks and stereo tracks, and you can use any of those tracks for input *or* output of audio signals. No matter the make or model, mixing boards are outrageously versatile. If you're podcasting with a friend, you can hook up two mics through a mixer so you won't have to huddle around the same microphone or slide it back and forth as you take turns speaking. Multiple microphones are the best option when you and your friends record. A real advantage of the mixer is that it allows you to adjust the audio levels of those multiple inputs independently so they sound even.

You may also be wondering about all those wacky knobs on a mixing board. Some of the knobs deal with various frequencies in your voice and can deepen, sharpen, or soften those qualities, and perhaps even help filter out surrounding *background noise* (which is the sound of an empty room because even in silence, there is noise). The knobs on the mixing board that are your primary concern are the ones that control your volume or *levels*, as the board labels them. The higher the level, the more input signal your voice gains when recording. If one of your tracks is being used for output, the level dictates how loud the playback through your headphones is.

Heavy-metal legends Spinal Tap may prefer sound equipment that "goes to 11," but cranking your mixing board way up and leaving it that way won't do your podcast much good. The best way to handle sound is to set your levels before podcasting. That's what's going on when you see roadies at a concert do a mic check. The oh-so-familiar "Check One, Check Two, Check-Check-Check . . ." is one way of setting your levels, but a better method is just rambling on as if you were podcasting and then adjusting your sound levels according to your recording application's volume unit (VU) meter. For more on setting levels, hop to Chapter 7 for all the details.

ANALOG VERSUS DIGITAL

So you decide to go mixer shopping and you see a lot of these mixers going for $40 or $50, and in some cases the same number of inputs at two to three times the cost. What gives?

Before rushing off and making a purchase you may regret later, understand the major types of mixers and the features they have. The two different types of mixers are *analog* and *digital*. What makes a digital mixer a digital mixer is how it processes the signals internally before it gets sent to your computer or recording device. Analog mixers have the benefit of being simple to set up, configure, and in the case of those with USB interfaces — yes, analog mixers can have digital outputs — easy to connect. The drawback is they are "hard wired" in their configuration and any additional signal processing needs to be done with additional hardware.

What kind of processing, you ask? *Compression* may be needed to bring up low spots and dampen loud spots, or you might need a *gate* which can set a *threshold* (a minimal noise level) before getting through to the mixer. These are nice features to help clean up the audio, but not mandatory. Digital mixers take the inputs from microphones, instruments, and other devices, and turn it in to a digital signal inside the mixer. They often have on-board compressors and gates for each channel and often complex input and output routing configuration.

Most podcasters don't need this level of sophistication, but Chuck Tomasi isn't most people. After his Tascam DM-4800 mixer gets done processing everything, the audio still comes out the back and connects to his Zoom H6 recorder with an analog jack.

In the case of mixers, you do get what you pay for. Cheaper mixers often have inferior audio output because of cheap pre-amps or other components. They often are poorly constructed and don't last as long. For something that sits on the desk, it's amazing how much abuse a mixer can take. When considering how much to spend, use the same rule with mixers as computers — get as much as you can afford. Having a professional microphone with a low-end mixer can put your overall quality at risk.

Hooking up a mixer to your computer

Now that you have your desk cleaned off (mind the dust bunnies!) and a perfect place for the mixer, set your mixer where you want it. And then make certain you can see the following items:

>> Your USB mixer

>> A power supply

>> A USB cable

TIP

Before hooking up the new mixer, check the manufacturer's website for any downloads (drivers, upgrades to firmware, patches, and so on) needed to make your digital mixing board work.

USB mixers are so similar in setup that we can give the same steps for both kinds of mixers. Regardless of where you fall in the Mac/PC debate, you can follow these steps to hook up your mixer board:

1. **Shut down your computer.**

2. **Connect the power supply to the back of the mixer and to an available wall socket or power strip.**

3. **Find an available USB port and plug your mixer-appropriate cable into the computer.**

   If your computer's ports are maxed out, we recommend investing in a PCI card that gives your computer additional USB ports. Do not run your digital mixer through a hub because that will affect the quality of the audio.

   A direct connection between your mixer and computer is the best.

REMEMBER

4. **Plug the USB cable into the back of your mixer.**

5. **Connect your input devices (microphones, headphones, monitors, and the like) to the mixer.**

6. **Power up your mixer by turning on the Main Power and the Phantom Power switches. See Figure 2-5.**

FIGURE 2-5:
Power switches for the board and for phantom power are located on different point of your mixer board, depending on the model.

7. **Start your computer.**

8. **Install any drivers your USB mixer may need.**

   Follow the instructions according to the manufacturer's enclosed documentation. Restart your computer if necessary.

And that's it! You're ready to record with your USB mixer. Now with headphones on your head and some toying around, you can set levels on your mixer good for recording.

TIP

Before filing away the reference manual that came with your mixer, be sure you know how things work. Buttons like the "Mix to Control Room" suddenly make sense to you as opposed to being "that button that needs to be down when I record." Getting a grasp of how things work on your mixer only makes you a better podcaster, so keep that reference manual close by and set aside a few pockets of time to clock in some it.

Accessorize! Accessorize! Accessorize!

A microphone (or microphones) and a mixing board serve as the foundation of your studio. You now need to add the final touches, as Martha Stewart would no doubt tell you if she were helping you with this process. When it comes to accessories, Martha might make suggestions like a doily for the mic stand or a sweet, hand-knitted cover for the mixer. When we talk about accessories, we have something different in mind. Here are some optional add-ons that can help you produce a rock solid podcast:

>> **Headphones:** Headphones help you monitor yourself as you speak. That may seem a little indulgent, but by hearing your voice, you can catch any odd trip-ups, slurred words, or missed pronunciations before playback.

TIP

If you are torn between making an investment in either the microphone or the headphones, we suggest you invest in quality headphones before you spend a lot of money on a better mic, or else you won't know how good a microphone is. Good headphones are nice to have for monitoring, but really show their value during the editing and mastering process.

Another advantage with headphones is they have better sound quality than your computer's speakers as well as reduce ambient noise around you if you purchase what are called *closed-ear headphones*. Keep it simple for your first time out with the *Behringer HPS3000* closed-ear headphones for around $30. These headphones end in a 1/8-inch jack but come with an adaptor that makes them a 1/4-inch stereo male jack (as shown in Figure 2-7), plugging directly into your mixing board.

Bose Noise Cancelling stereo headphones are going for $300 and promise the highest quality for audio listening. Note that we say listening — not recording and editing. Any headphones you see listed as noise-cancelling are going to be terrific for listening to audio, but those headphones aren't the best for the recording and editing process. You want to be able to hear any noise that these headphones cancel out so that you can eliminate it before a word is recorded. The best noise reduction and elimination happens before and during recording, not afterward. For podcast production, stick with closed-ear headphones sans the noise-cancelling feature.

» **Cables:** As mentioned earlier in the chapter, your newfangled microphone may arrive without any cables, and buying the wrong cable can be easy if you don't know jack (so to speak). So check the mic's connector before you buy.

With many high-end mics, the connectors aren't the typical RCA prongs or the 1/4-inch jack shown in Figure 2-6; instead, you use a three-prong connection: a 3-pin XLR plug, as shown in Figure 2-7.

FIGURE 2-6:
A 1/4-inch male connection, which is needed to connect headphones to a mixing board.

FIGURE 2-7:
XLR male and female plugs, standard plugs for phantom-powered microphones.

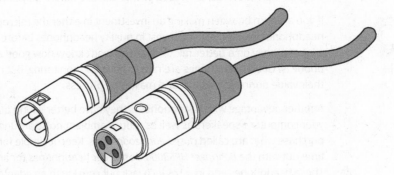

To plug a microphone into your mixer, you want to specify a *3-pin XLR-to-XLR male-to-female* cord; the female end connects to the mic, the male end to the mixer. These cables begin at $9 and work their way up, depending on the length of the cord and quality of the inputs.

>> **Microphone stands:** On receiving the microphone and possibly its shock mount, you may notice the attachment for a mic stand . . . but no mic stand comes with your new mic. It's your responsibility to provide one, and although that may sound like an easy buy, your options for mic stands are many, each with their advantages.

A simple *desktop mic stand* can run you around $10 and is the most basic of setups. When shopping for the right height and make, you'll notice other stands like a *boom mic stand* around the $100 range. The type of mic stand that is best for you depends on what you want it to do and how you want to work around it. With the inexpensive desktop stand, you're ready to go without the extra hassle of positioning and securing a boom stand to your desk. The boom stand, though, frees up space on your desk, allowing for show notes and extra space for you to record and mix in. Consult your budget and see what works for you.

>> **Pop filters and windscreens:** Go to any music store and ask for *pop filters* and *windscreens,* which are both shown in Figure 2-8. Both devices can help soften explosive consonants (percussive ones like *B* and *P*) during a recording session. A windscreen can also reduce some ambient room noise. Using both on one mic could be overkill, but these are terrific add-ons to your microphones.

FIGURE 2-8: Windscreens (left) and pop filters (right) help you control unwanted noise and percussive consonants (such as *P* and *B*) during recording.

WARNING

It is worth mentioning that there are DIY tutorials on YouTube, blogs, and elsewhere that will show you how to make pop filters with a stick and panty hose. Just remember: You get what you pay for. With the time, experimentation, and jury-rigging in order to save yourself $10-$20, it's less stress for you to just go and pick up a proper pop filter.

RODECASTER PRO — THE ALL-IN-ONE PACKAGE

If you are interested in simplifying the setup of mixer, mics, and many of the accessories, here's a tool to consider — the Rødecaster Pro (https://www.rode.com/rodecasterpro). This isn't just a mixer; it combines a mixer with soundboard, recorder, headphone amp, and much more. The four XLR inputs are great when you and your friends want to get together in-person. Additionally, it includes a USB port, which serves two purposes. First, the Rødecaster quickly plugs in to a laptop, so apps like Skype can use its USB connection as the mic and speakers.

Second, the USB port can be used with the free companion app to transfer your pod- cast files from the micro SD card (sold separately) or upload audio clips to your sound- board. And one of Chuck's favorite features is he can pair his mobile phone via Bluetooth for easy interviews! The right side sports eight large sound effects buttons. Using multiple screens, each button can have a variety of functions and color coding. This makes it easy to remember which button is your intro, outro, interview, or unused. The unit even includes four headphone jacks, each with individual volume controls for your guests. How does the Rødecaster Pro sound? Great! Chuck complemented the device with a Røde Pod Mic and was impressed. For those who want to go deeper, there are advanced settings to change the onboard equalizer, compressor, and gate. The base unit lists at $599. You also need to add mics, cables, SD card, and headphones. The value of such an investment is the simplicity and quality.

Chuck's only regret: The Rodecaster Pro was not around when he started podcasting back in 2004.

Chapter **3**

Building Your Podcast's Digital Workstation

After you have your recording equipment in place, plugged in, and running, it's time to take a look at *audio-editing* software packages. These applications help you take that block of audio marble and chisel the podcast hidden within it.

REMEMBER

If you are looking to podcast video, we cover streaming content and how it fits into the podcasting workflow in Chapter 17.

As with digital photo editors, video-editing software, and word processors, *digital audio workstations* (DAWs) come in all sizes and all costs, ranging from free to roughly an entire paycheck (or three). Like any software package, the lower the cost, the simpler the product and the easier it is to understand, navigate, and use for recording. However, as the software grows more complex (and expensive), the features that offer professional-level results become abundantly clear. In this chapter, we run down some of the audio recording/editing software that may be right for you.

TIP

When you have software and hardware in place, test your setup to make sure everything works. Take a look at your application's preferences so that your sound input and output are going through your mixer, bring up the volume on the channel on the mixing board that your microphone is connected to, and then listen to yourself through your headphones. Just make sure everything is running as expected so you can jump right in and get recording.

Budget-Friendly (a.k.a. Free) Software

Whoever said "You can't get something for nothing . . ." didn't know about podcasting. It's amazing what kind of production you can create with little or no monetary investment.

Audacity: The risk-free option for all

Audacity (shown in Figure 3-1) is a piece of software that quickly became a podcaster's best friend. It's easy to see why: It's free, simple to use, and safe to download. It's available at `https://www.audacityteam.org` and provides a common starting point for new podcasters.

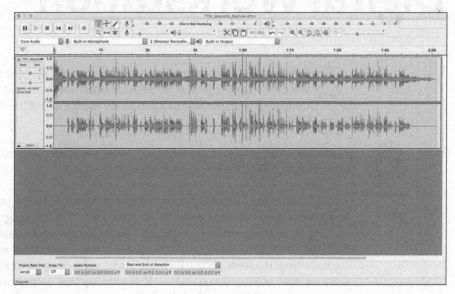

FIGURE 3-1:
Audacity is an open source, audio recording, and editing application that allows you to edit audio and create MP3 files.

Designed by the open source community who simply wanted to "give back to the Internet" something cool, Audacity is designed for a variety of audio capabilities such as importing, mixing, editing, and exporting audio and has earned a reputation for being the must-have tool for recording voice straight off a computer. It's also compatible with Windows, Macintosh, and Linux. (We want to send a big thank you to the volunteers who went out of their way to show that yes, software can be made available for any platform, provided the creators are driven enough to make it happen!)

>> Records live audio through microphones or mixer channels

>> Can record up to 16 channels at once

>> Imports various sound formats for editing and remixing

>> Exports final projects to WAV, AIFF, MP3s, and many other audio formats

>> Grants the user unlimited Undo and Redo commands

>> Can create an unlimited number of audio tracks

>> Removes static, hiss, hum, and other constant background noises

>> Offers a wide variety of effects to manipulate your audio (and these effects are expandable via third-party plug-ins)

>> Records at an audio quality of up to 384 kHz

Cakewalk by Bandlab for Windows: A complementary Step in Running with the Pros

Audacity is an excellent piece of software for the basics, but what if you desire more control over the capabilities and features of your audio-editing package? Are you looking for more recording options, additional audio filters, built-in multi-track recording, and pre-recorded music loops? (Gads, what some people will do to set a mood or a tone for a podcast.)

Cakewalk by Bandlab (https://www.bandlab.com/products/cakewalk), formerly known as Cakewalk's SONAR, is a free download of a fully loaded DAW exclusively for Windows.

Cakewalk (shown in Figure 3-2) offers its users some serious goodies:

>> Record audio from a microphone or other media (CD, LP, cassette, or Internet audio stream).

>> Display and edit beat-by-beat with audio waveform displays.

>> Pro-grade audio engineering tools included: Chorus/Flanger, Compressor/Gate, Tempo Delay, Modifier, Reverb, and other post-production effects.

>> Support for MP3, WAV, and WMA files.

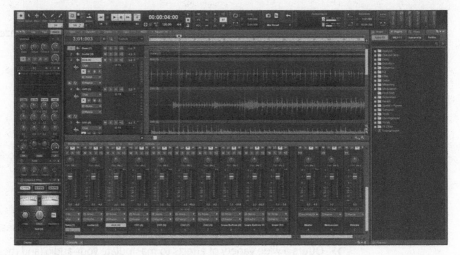

Cakewalk allows users to record, edit, mix, and produce audio compositions, whether it is an original score for your podcast, a continuous music mix, or just you and your own personal soundtrack. Between its easy audio-editing features and Cakewalk's collection of original audio samples (called *Cakewalk Loopmasters Content Collection*), it's a breeze to set your own themes and special effects.

GarageBand: Moby in your Mac!

Cakewalk by Bandlab is another free audio package for the Windows user looking to go beyond Audacity, but what about Mac users? For people still in the minority of the computer world, it's always a frustration to hear software developers say, "No, we won't be making this product available for Mac users." Sometimes Mac users seem to be denied the coolest toys and utilities because they just aren't offered for reasons not revealed. (Maybe a penalty for thinking differently, but still . . . .)

Apple's creative crew understands this injustice and thusly came a software gem in 2004 that made it more than cool to be a Mac user, especially one who's into podcasting. Since the early days of the iPod to today, Mac-using podcasters continue to create audio productions with it — *GarageBand*.

With hundreds of music loops that can easily switch from one instrument to another, GarageBand (shown in Figure 3-3) makes royalty-free music easy to compose, special effects a breeze to create, and podcast episodes effortless from conception to content. Many of the loops are editable and, with a bit of tweaking, can set the right mood for your podcast. GarageBand has a few new additions:

- » Multitrack recording supports simultaneous tracks recording independent audio sources all in one recording session.

- » With the Multi-Take feature, you can now do multiple readings or segments and save them as separate files. Then you can pick which one you want for the final.

- » You can see a display of the music, with actual notation, in real time.

- » You can output your audio directly to Apple Music, directly to SoundCloud for distribution, or locally on your drive for additional post-production or archiving.

- » You can save the recordings as loops in the GarageBand library.

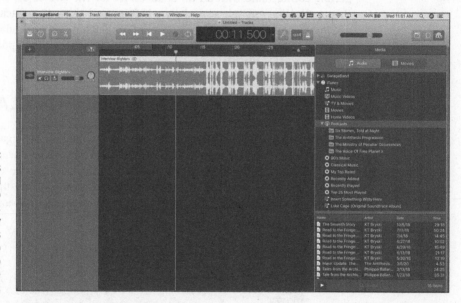

FIGURE 3-3:
GarageBand is easy to use and even easier to have a blast with, as a large library of sound loops and access to your own audio libraries are made accessible.

But perhaps the most amazing aspect of GarageBand, as reported by *Ars Technica* at https://bit.ly/AT2017-GB back in 2017, is that GarageBand is now free for anyone using a Mac OS or iOS device.

GarageBand's most appealing asset (apart from the fact that it is free now) is its hundreds of sampled instruments available in loops. You can easily edit and splice together these loops with other loops to create original music beds of whatever length you choose. Prerecorded beds range from Asian drum ensembles to Norwegian Folk Fiddles to Blues Harmonica to Emotional Piano reminiscent of films like *The Fault in Our Stars* and *Sense and Sensibility*.

GarageBand also provides a capability — with many (not all) of the samplings — to create your own musical theme. Sure, some instruments may sound better than others, but you might — with a bit of trial and error — create an original melody that serves as the best royalty-free intro and exit for your podcast.

WARNING

If you're planning to do a bit of composing in GarageBand, be warned. Garage-Band is a lot of fun but can easily soak up free time that you would normally dedicate to recording. So if you need to get in touch with your inner Mozart, set aside a good-size pocket of time to put together your desired riffs. You have many options to choose from; as with podcasting, it's best not to rush the process.

GarageBand is easy to navigate and understand (no, not master, but definitely understand) within a short span of time. Plenty of terrific books are available for getting comfortable with all its nuances. As for the two of us, we cannot praise this application enough, especially with so many expansion packs available that add instruments, riffs, and loops to your GarageBand (always updated and stocked at the App Store). This unassuming software offers a lot to the podcaster.

SHARING WITH THE REST OF THE CLASS

We're particularly keen on GarageBand, which gives you the option to organize your final audio files in an Apple Music playlist. When you're done with that particular recording session (what GarageBand refers to as a "song"), choose Share ➪ Song to Music. Before your final audio is exported, you are offered options to tag your audio with a variety of credits, including your own playlist. Once exporting concludes, Music opens with your new playlist. See Chapter 9 for details on adding ID3 tags.

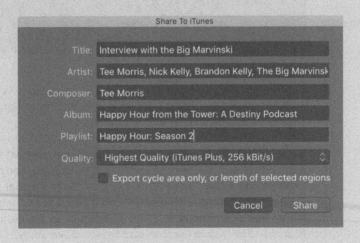

The Sky's the Limit: Big-Budget Software

If you're lucky enough to have unlimited funds and resources to build your podcasting studio, this section on software is for you. A majority of podcasts are working on the bare-bones plan, and so far the investment in the equipment we've recommended (in Chapter 2) is for a budget of under $500 — provided you feel like making an investment in a professional microphone, mixing board, or software.

REMEMBER

You may hear podcasters say "Content is king," meaning it doesn't matter how good you sound if you don't have anything to say. What makes a good podcast is the same whether you're on a budget of $0, $500, or $Ridiculous.

The difference is in the sound you can get. For the corporate entity, government agency, or professional organization venturing into podcasting, sound quality is crucial as your reputation and experience are now being "socially tested." Do you not bother with the details, or do you raise the quality bar? Commercial podcasting demands nothing less than the best in audio quality, and that is what investments in professional software bring to your production. From noise reduction to production parameters required for services like Audible and Spotify, high-end audio software gives you full control over every aspect of the audio you're recording.

Adobe Audition

At one time, a favorite software application was CoolEdit. But when Adobe Systems purchased it and repackaged the software as Adobe Audition (shown in Figure 3-4), it got even better. Audition (www.adobe.com/products/audition.html) is offered as a standalone for $20.99 per month or as part of the Adobe Creative Cloud membership for $52.99 per month. Audition's features are nothing short of awesome:

>> 128 tracks available to the user

>> Can record 32 different sources simultaneously

>> Offers 50 digital sound effects to enhance your audio tracks

>> Provides 5,000 royalty-free loops that can be easily edited and compiled to create your own music beds

>> Offers Surround Encoder for 5.1 surround sound mixing for audio only or integration into an Adobe Premiere Pro project

>> Offers audio restoration tools like Click/Pop Eliminators and Noise Reduction that can restore recordings from vinyl and cassette recordings; remedy pops, hisses, hums; and fix clipped audio

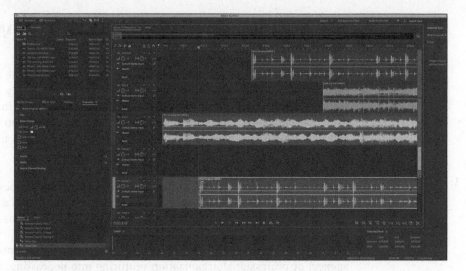

FIGURE 3-4:
Adobe Audition is the professional standard software for editing and engineering audio that gives the podcaster complete control.

What makes Audition so appealing to professionally engineered productions such as P.G. Holyfield's epic *Murder at Avedon Hill* (`https://bit.ly/Holyfield-MAAH`) and the ongoing *Secret World Chronicle* (`http://secretworldchronicle.com`) is how it gives you dominion over pitch, wavelength, time-stretching, background-noise removal, and Dolby 5.1 stereo output, making this DAW a staple in the digital audio industry.

Audition runs on Windows and Macintosh platforms. If you know your podcast needs a professional polish and you are ready to make the investment and the jump to higher grade software, Audition may be the option for you.

Apple Logic Pro X

Apple Logic Pro X (`http://www.apple.com/logic-pro/`) is another software package in the industry that is carving out a place for itself with professionals. Designed to be the next step after GarageBand, Logic Pro X (shown in Figure 3-5) has been given features built for Mac-based podcasters:

>> Supports multitrack recording.

>> Displays multiple audio takes in a single window. After you select your desired takes, Logic Pro X compiles them for a final composite — complete with *crossfades* (the simultaneous fading from one audio source into another) — creating seamless playback for the end result.

>> Copies audio effects and EQ settings from another clip — setting them on an audio clipboard for quick access and application or saving them as a preset for future use.

FIGURE 3-5:
Apple Logic Pro X
is a fantastic tool
for podcasters
aiming for a
professional
polish to their
production.

» Features specific functions for producing podcasts, such as ID3 tagging.

» Outputs multitrack audio projects into 5.1 surround sound.

» Offers multiple audio formats for export including WAV, AIFF, and many others.

Apple Logic Pro X offers incredible options for the audio professional and the up-and-coming podcaster. For $200USD from the App Store, you can accomplish a lot with it. Tee does, and swears by it.

Gluing It Together with RSS

The hardware (mics and mixers) and the software (GarageBand and Audacity) are necessary to record audio and create the podcast media file. That's the fun and creative part. But to make your recording a podcast, you need to get your hands dirty on the tedious and technical parts and add one more three-letter acronym to your vocabulary: *RSS.*

Since the first edition in 2005, we've helped a huge number of podcasters get started — and in nearly every case, the RSS step is the biggest source of confusion. So, there's a lot to RSS, which we can explain . . .

XML (eXtensible Markup Language) is a *markup language* like HTML (the building blocks of the Internet), and RSS (Really Simple Syndication) is a file format built on XML. XML's purpose is to allow systems such as computers, network devices, and other gadgets to exchange information in a structured format. The RSS file is that bit of information that the podcaster publishes so others can use their pod-catching software to check for new content automatically.

If that somewhat technical explanation doesn't do it for you, try this old school analogy about deliverable content. Consider journalists for publications like *The Washington Post, The New York Times, Vanity Fair*, or *Wired Magazine* as authorities on what is happening within your passions in life. Whether the subject matter covers Apple computers, the Boston Red Sox, or The Beatles, you know these journalists are trusted experts in their fields, or maybe they are fellow fans just as passionate as you are. These journalists, if they are truly worth the ink used in their articles, are always creating new and unique content with every new issue or publication.

What if *you* want that content? Well, you got options. You can check the newsstand daily. Maybe it's there, or maybe it is sold out on that particular day so you come back another time to see if the latest issue is there. Another way to get that content is to subscribe to the magazine or newspaper, and the content is delivered to your doorstep.

Here the role of the magazine is filled by the media created by the podcaster, and the RSS is the delivery mechanism that sends your podcast to people who *subscribe*. Just like you would with a magazine or newspaper.

No school like the old school, huh?

SPLITTING HAIRS: WHAT DEFINES A PODCAST

Regardless of this or other books published on podcasting, there are people who think if media is posted on a blog or website and a "Podcast" graphic is placed over it, then you are podcasting media. There is also the notion that you record for your friends, a snappy show about a favorite topic, post it on YouTube, and you have a podcast. The tricky thing about this ongoing debate is that podcasting has evolved since we first started. Streaming content can be consumed on mobile devices, laptops, and desktop computers and then circulated through blogs, communication apps like Discord, and various social media platforms. The lines are blurring as to what defines a podcast, but for the sake of this book, we are defining a *podcast* as a delivery of media from your studio to a consumer using RSS. A podcast happens when you have these three elements in place — a website or blog, some form of media, and an accompanying RSS feed to deliver said media.

IT'S ALL ABOUT THE <ENCLOSURE>, BABY

If you already have a blog, you're already generating an RSS feed. Although podcast apps like Overcast, Stitcher, or Spotify can read your RSS feed, the feed needs to include the <enclosure> tag in order for podcasting to work.

Dave Winer invented the <enclosure> tag in early 2001 for the purpose of embedding links to large audio, video, or other "rich media" elements into an RSS feed. At the time, Dave and Adam Curry were trying to solve the click-and-wait problems inherent in big files such as audio and video. Back then, if a user clicked a link to a 30MB file, several minutes would drag by before the file was completely downloaded to the user's hard drive and was usable. Not a good user experience, regardless of what's in the file.

With the advent of the <enclosure> RSS element, users could subscribe to places where they expected large files as a regular occurrence and move the downloading of those files to the early hours of the morning, when the users were snug in bed and a ten-minute download was no big deal.

Of course, users back then had to be technically savvy to take advantage of this new RSS element. It wasn't until the summer of 2004 that Adam Curry wrote what most consider the first podcatching client — a simple, user-friendly desktop program that extracted enclosed media files from RSS 2.0 feeds. And behold! Podcasting was born.

As a podcaster, your job is to make sure you keep the RSS updated and current each time you post a new podcast media file. Lucky for you (and us too), plenty of software solutions make this step a breeze.

Keep it simple and get a blog!

If you're looking to spend the least amount of time hand-coding an RSS feed, look no further than starting up a blog. Easy to set up and often free of charge, blogs make the process of generating and updating RSS feeds insanely easy by doing it automatically.

You can choose from dozens of blog software packages (also called *engines*), each with a variety of bells and whistles that are designed to make your updates (including your RSS feed) as easy and/or customizable as possible. For a crash course in how blogs do what they do, check out *Blogging For Dummies*, 7th Edition by Amy Lupold Bair. Her book can help you choose which blog engine might be right for you. Meanwhile, here are a few options we prefer:

>> **WordPress** (`http://wordpress.org`): Perhaps the most popular blogging solution in the writers' eyes, WordPress has many advantages over other free blogs. Not only is it easy to install and get running, but it also supports podcasting out of the box. When you incorporate the PowerPress plug-in (available at `https://wordpress.org/plugins/powerpress/`), podcasting takes on a whole new level of "user friendly" options. There are also thousands of WordPress and user-developed templates that can be used as is or customized to fit your specific look and feel for your podcast. Oh, and it's free. To make incorporation into your website seamless, many hosting companies like DreamHost (`http://dreamhost.com`) and GoDaddy (`http://www.godaddy.com`) offer packages that have WordPress preinstalled or installed with a one-click option. Figure 3-6 shows the WordPress signup page.

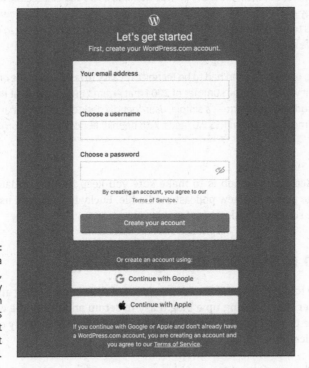

FIGURE 3-6: WordPress is a popular, user-friendly blogging option that helps podcasters get their shows out into the world.

If you choose to go the WordPress route, pay particular attention to the web address. There's Wordpress.org and Wordpress.com. The .org flavor is where you can download the software and install it on your server of choice — this option involves a bit of technical work. The .com site has the software already installed and you just configure it; however, the configuration options are a bit more limited.

There are two basic options regarding where to put your blog, and prices and products vary from vendor to vendor. The *nonhosted model* blog is where you get to choose the blog engine, a great option if there is a specific engine you want to use. However, there is more work involved with setting it up and maintaining it. The other option, the *hosted model*, keeps things simple. The choice of blog engine is made for you. The trade-off for flexibility is ease-of-use.

» **LibSyn** (`http://libsyn.com`): Don't be surprised if Liberated Syndication turns up quite a lot in this book. LibSyn, still going strong after launching in 2004, is a combined blog/hosting company specifically designed for podcasting. Although it may not address all your needs, its ease of use and all-in-one nature should not be dismissed. Some podcasters choose to use LibSyn as their hosted solution; others use it to store their podcast files while their blog is on a non-hosted solution somewhere else. Either way, we think LibSyn's pricing is quite reasonable, starting at $5 per month.

» **Podbean** (`www.podbean.com`): Another popular all-in-one option with the added support of a podcast directory optimized for mobile devices, Podbean combines online hosting packages, podcast publishing tools, and crowdfunding services. Podbean, both in its website and mobile app, offers a user-friendly interface integrating feed and media management, syndication, and analytics, along with promotion through popular social media platforms like Facebook and Twitter. All this is available through affordable, flat-rate hosting plans. If you are starting off with baby steps, and then looking to take a deeper dive into podcasting, then Podbean is the right option for you and your podcast.

Doing it by hand

What's this about podcasting by hand? Are you crazy? No, seriously — generating an XML file in RSS 2.0 format isn't overly difficult. It's just extremely easy to mess up! Sure, you could open Notepad, TextEdit, or any text editor, download a few examples, and generate your own code, but we advise against it.

However, some code warriors — in the same vein of hand-coding HTML, JavaScript, and other markup languages — insist on composing their own code from the ground up to this day. You glutton for punishment, you.

If you decide to run your podcast on a home-spun, handmade RSS feed, you're going to want to make sure your feed is solid and ready for prime time. That's when *Podbase* is your best friend and harshest critic. Go to `https://podba.se/validate` and enter your feed's URL (not the host blog or site, but the *feed* itself) and then click the Validate button. Podbase tells you whether your RSS is air-tight.

You can also find the full technical specifications for RSS 2.0 and a few answers to questions you have about RSS. This is a great place to visit not only when you're about to launch a podcast but also when your feed suddenly stops working. (This is also a good "In case of emergency . . ." site for when you have WordPress handling the heavy lifting, and one day throws its back out.) When something is borked in your podcast, a feed validator is the first place you want to visit.

Podcast Management 101

Unless you already have hosting taken care of, you're going to need a place on the web to put your stuff. You know — your podcast media files, RSS feed, and show notes for your podcast. You also need a way to get them up there.

Getting a hosting provider is a breeze, with hundreds of companies all vying for your precious, hard-earned money each month. The good news is that all this competition has brought down the cost of hosting packages significantly. The bad news is that you have to go through a lot of clutter to reach the right selection.

This section covers the basic needs for most beginning podcasts and mentions a few pitfalls to watch out for. In Chapter 10, we get into the process of actually moving your files to your host.

TIP

Don't rush into a hosting agreement just yet. We suggest reading the rest of this chapter as well as Chapter 10 before forking over your credit card. We cover lots of good information that can help you narrow your choices.

When you're comparing hosting plans, try not to get bogged down in the number of email addresses, MySQL databases, subdomains, and the like. All of those features have their own purposes, but as a podcaster, you have only two worries: how many podcasts the site can hold and how much bandwidth you get.

Size does matter

Podcast media files are big. Unlike bloggers, podcasters eat up server space. Where simple text files and a few images take up a relatively small space, podcast media files tend to be in the 5MB to 50MB range. And that's just for audio.

Here some suggestions for zeroing in on what you need storage-space-wise:

>> Think about how many podcasts you want to keep online and plan accordingly. If you have a plan that isn't constrained by storage space, then you don't need to worry about it.

>> Consider the amount of server space you'll need to host your blogging software, databases, text, and image files. For example, if Tee wants to keep all episodes of *The Shared Desk, The Minstry of Peculiar Occurrences,* and *Happy Hour from the Tower* online, he will easily need several gigabytes of space, as one of these shows first went online in 2010. Podcasting did help quite a bit in creating a need for server space.

Podcasters should look for hosting plans that include at least 3GB of storage space. As of this writing, several host providers charge less than $10 per month for that much space, and more.

Bandwidth demystified

Of equal importance to storage space is *bandwidth,* an elusive and often-misunderstood attribute of web hosting that is critical to podcasters. Bandwidth refers to the online space needed to handle the amount of stuff you push out of your website every month. The bigger the files, the more bandwidth consumed. Compounding the problem, the more requests for the files, the more bandwidth consumed.

For instance, the bandwidth for Chuck Tomasi's *Technorama* can be over 100GB a month — that's pretty impressive. Why so huge? Chuck and his cohost Kreg just celebrated their 600th episode. They've been at podcasting since the very beginning, and at their initial launch, the amount of information exchanged (read: downloads) was modest, but then people started talking. With the rise in popularity (thanks to top-shelf interviews with people in the science and entertainment industries), their downloads increased. So did the demands on bandwidth. Web hosts, in situations like this, must consider when it's time to allocate a larger bandwidth package to handle a show — and that means more cash outlay.

And therein lies the double-edged sword of success. Most podcasters want more listeners, and that means more podcatching clients requesting the podcast media files. Bottom line: The more popular your show gets, the more bandwidth is being consumed every month.

To simplify, pretend that you produce one show each week, and your show requires 10MB of bandwidth. You publish the show on Monday, and your 100 subscribers receive your show that evening. You've just consumed 1,000MB of bandwidth (100 × 10MB) for that week. But next week, more people have found out about your incredibly amazing show, and now you have 200 subscribers. Next Monday, your bandwidth increases to 2,000MB, which gets added to your previous week's total to bring you up to 3,000MB. The new listeners were so happy, they also download the previous week's show, tacking on an extra 1,000MB. You've just consumed 4,000MB (or 4GB) of bandwidth for the month. You still have two weeks to go in the month, and if your numbers continue to climb like this, you will be burning through bandwidth (and your web host budget) quickly!

As a general rule, the longer your podcast episodes are, the more bandwidth you will need. If you find a plan that offers unlimited bandwidth, the problem is solved. Otherwise, you'll want to try and find a plan that's high in the gigabytes. If you have a podcast that's both long and popular, start looking at plans that offer 1TB (that's *terabytes*) or more.

You have ways to avoid the issue of bandwidth, or at least make it less of a concern even if you have large ambitions. Chapter 10 talks about some podcast-specific and advanced hosting options. Even if you don't think you'll have to worry about bandwidth, it's a section to pay close attention to because you'll likely be more popular and perhaps *wordier* than you think.

Chapter **4**

Go, Go, Power Podcasters!

U p to this point, we have been building our podcast studio with your home computer as the center of your digial audio workstation. We have a mixer plugged with a few mics, and we have our software up and running. Or maybe we have a USB mic hooked up to your desktop computer (directly, and not through a USB hub, remember?), and you have Audacity primed and ready to go. Welcome to your new audio recording studio.

But hold on — what if we want to record someplace else? What do we need to do to pack up the whole recording unit and take this show on the road? What if your podcast isn't about where you are but where you are going?

That's why we're spending some quality time on *portability* in this chapter. The aim is to be able to pack up our podcast and set up wherever we stop. Yes, yes, yes, we know — podcasts are portable by nature, but we're talking about packing up the studio and working on location. In this chapter, we're taking the studio to go. Super-sized. And a happy-cool-fun meal for the kids.

TIP

When you have software and hardware in place, especially when you're working remotely, test your setup to make sure everything works. Take a look at your System Preferences to make sure you are working with the proper sound input and output sources, make certain your levels are solid, and then run a quick sound check through your headphones. It is a good idea to make sure everything is running as expected so you can jump right in and get recording.

Podcasting with Your Laptop

So you want to have a podcast that has that studio quality sound, but you want more than one microphone, and be able to have a bit of control over the levels. This is when you would want to invest in a preamplifer or a *preamp*. To understand what a preamp is, you should learn a few more technical matters around microphones.

Microphones, be they condenser or dynamic, record their signals at *mic-level*. This is the signal created from the internal diaphragm moving back and forth against a magnet in a wire coil, generating an electrical signal. It's a clean signal, but very weak. The best audio is recorded not at mic-level, but at *line-level*. You get line-level signals coming out of electric guitars, keyboards, and other instruments. To get a weaker mic-level signal boosted to the line-level signal, you need to give it a swift kick-in-the-pants. The preamp, sometimes a separate unit or built into a mixer board or a USB microphone, provides that kick to bring the mic-level signal to line-level without adding any noise to the original signal.

Now that you know what a preamp is, how about a few options for you to consider?

Mackie Onyx Blackjack

The *Mackie Onyx Blackjack* offers you all the power and reliability of a mixer in a small, compact design. The best part of working with this preamp is it has zero latency when recording. This means there is no delay for when you speak and when you hear your voice while recording. You can adjust the Blackjack's buffer settings to maximize your computer's processing ability as well.

Blackjack also offers podcasters:

>> A prepamp bus-powered via USB

>> Two XLR connections delivering 48V phantom power

>> A 25-degree inclination by design, allowing for full view of all controls at all times

>> An all-metal chassis that gives the Blackjack "built-like-a-tank" durability

>> Onboard analog-to-digital conversion, granting your amplified signal with the lowest noise and distortion possible

Shure MVi

Shure Audio, a manufacturer we name in Chapter 2, is no stranger to setting the bar for audio engineering and recording on looking at the prevalence and relevance of its audio gear. With the rise of podcasting, Shure set out to create gear that would capture quality sound, and the *MVi* is a compact, USB-powered preamp ready to power your microphone (or microphones, if you employ a splitter) accordingly.

The MVi, shown in Figure 4-1, offers a podcaster-on-the-go:

>> USB connectivity for easy plug-and-play with optional iOS connectivity with iPhones and iPads

>> Touch-sensetive panel for control over five different DSP presets, headphone volume, and more

>> Built-in headphone jack for real-time monitoring

>> One XLR connection offering 48V phantom power option

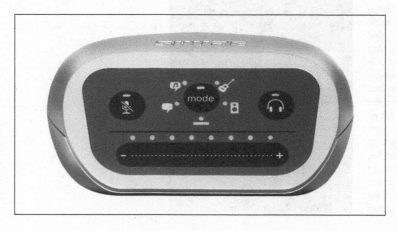

FIGURE 4-1:
Shure's MVi, part of the MOTIV series of audio gear, is a preamp designed with desktop computers and mobile devices in mind.

TECHNICAL STUFF

If you want to record on location, you could use studio condenser mics which, on account of their sensitivity, will pick up a lot of the background, setting a nice ambiance for your podcast. Depending on your environment, though, there might be too much ambiance for your interview. This is why, in most on-location settings, *dynamic* mics are preferred. You will still get some background noise, but not as pronounced when using studio condenser mics.

The Shure MVi can serve as a preamp, powered by USB, similar to the Onyx Blackjack. With an even more compact design, the MVi makes your portable studio even more so.

Podcasting with Your Mobile Devices

Development of mobile devices — smartphones like the Apple iPhone and Samsung Galaxy and tablets like the iPad and Surface — has been astounding to watch over the past decade. With cloud computing becoming more and more prevalent not only in personal lifestyles but now in corporate sectors, mobile devices big and small are quickly becoming the new alternative to portable computing. When once it would require an entire backpack and pockets of accessories to replicate the office, mobile devices have reduced portability to a single pocket and a connection to the Internet (see Figure 4-2).

FIGURE 4-2: Nobilis Reed of *Nobilis Erotica* and *This Kaiju Life* has a portable recording rig with a USB microphone, a smartphone, and Dropbox. His sessions are all sent to cloud storage (relieving demands on the phone) and are easily accessed by his editing computer.

Now, leading names in microphone technology are releasing new audio gear designed for the portability you come to expect from smartphones and tablets.

TECHNICAL STUFF

The figures featured throughout this chapter appear somewhat Mac-centric but do not affect the ability of these innovations to oat easily from platform-to- platform. What we showcase here, equipment-wise, does not discriminate from one operating system to another. It's all in the ability to connect.

A Shure Thing: The MV5 and MV51

Shure Audio's MOTIV series not only includes a preamp but also a new line of microphones built to be portable, and still able to yield superior audio recording quality Shure is known for.

The MV5

The *MV5* (shown in Figure 4-3) packs a lot of audio punch in its small, sleek, retro design and its economic cost of just under $100USD. Offering you professional-quality audio when recording, the MV5 provides latency-free headphone monitoring and quick-and-easy plug-and-play capability.

Other features of the MV5 include:

>> Retro-design includes desktop stand and adjustable position for microphone

>> Three digital signal processor (or DSP) presets

>> iOS and USB connectivity

>> Built-in headphone jack and volume adjustment for real-time monitoring

FIGURE 4-3:
Shure's MV5 (left) and MV51 (right), part of the new MOTIV series, offers podcasters high quality audio recording using any mobile device.

The MV51

The *MV51* (also shown in Figure 4-3) is much like its small counterpart, the MV5, in its retro design, ability to capture professional-quality audio, and zero latency when live monitoring. Beyond that, the MV51 is a fantastic addition to your ultra-portable studio on account of the additional features found only in this higher model:

>> Quick-and-easy plug-and-play capability

>> Retro-design includes pull out stand, either for standing independently on a flat surface or fitting into a standard microphone stand

>> Touch-sensetive panel for control over five different DSP presets, headphone volume, and more

>> iOS and USB connectivity

>> Built-in headphone jack for real-time monitoring

>> Large-diaphragm condenser capsule offering wider audio range for recording

The MV51 is more than just a step up from the MV5. It promises to set a new standard in mobile recording. With its capability to capture clean audio, setup and recording is as simple as unlocking your mobile device and recording on your audio app of choice. You are also given advanced options such as either recording with one of the MV51's onboard presets or simply recording at, uncompressed, unaltered audio, offering you full control over post-production treatment. Whether in-studio or on-the-road, at $200 the Shure MV51 proves itself a valuable asset in your mobile studio setup.

Two for the Røde: The VideoMic Me and smartLav+

Røde Microphones (http://en.rode.com/) have made a name for themselves since podcasting's early days. New Zealand's first podcast author, Pip Ballantine (http://pjballantine.com), won the 2009 Sir Julius Voguel Award for her podcast *Chasing the Bard*, recording her epic fantasy on the Røde Podcaster (http://en.rode.com/microphones/podcaster). Pip still endorses the USB microphone, suggesting it to everyone building a new home studio on a tight budget.

Røde continues to provide a complete array of mobile options for podcasters ready to take their recording out of the studio, and go one step further in specializing gear for smartphones. Why smartphones? Particularly in interview siutations where sitting is not an option (press conferences, man-on-the-street interviews, and so on), you are going to need your recording rig to be even more compact than a tablet. That is where your smartphone transforms faster than an Autobot (or a Decepticon, depending on the mobile OS you prefer) to become your recording studio.

VideoMic Me

Røde's *VideoMic Me* (captured in action at Figure 4-4) at the cost of just under $60USD upgrades your on-board smartphone microphone to higher-quality audio. Instead of that tinny quality found in most phone conversations and smartphone recordings, the VideoMic Me picks up a far wider range and frequency, while focusing the direction of the mic, yielding audio of a much higher quality. The accessory also comes with an optional mount for better mic stability and a windscreen to cut down on any unexpected weather elements you might encounter.

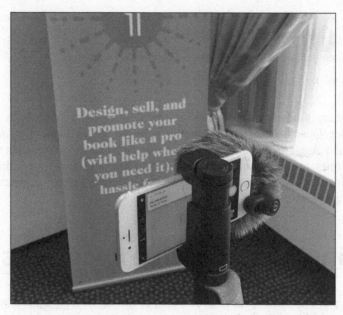

FIGURE 4-4: Røde's VideoMic Me transforms your smartphone into a handheld recording device (as seen recording on location at a convention), capturing higher quality audio than what an onboard microphone would normally capture.

WARNING

In turning smartphones into portable recording devices, you may find your recording riddled with intermittent static. This is *RFI* or *Radio Frequency Interference*. RFI occurs when a disturbance in frequency is generated by an external source — for example, other nearby electronic equipment — causing a degradation of a signal. Static. You can try recording in Airplane Mode, but this prevents you from doing any sort of mobile streaming, and even then, it is no guarantee. As always, test your mobile rig before trying to capture that once in a lifetime recording!

smartLav+

For the podcast more about the host or for the interview where the interview host is edited out of the final show and only the subject's voice remains, the *smartLav+* at a cost similar to that of the VideoMic Me offers broadcast-quality audio for a modest investment. The smartLav+ is a *lavalier mic,* meaning it is not held by or pointed at the subject but worn on lapels or the collar of a host or guest. Unlike other lavalier mics that are connected to a wireless transmitter, the smartLav+ connects directly into a smartphone or tablet headset jack and records using GarageBand, Røde's own Rec app, or any other media-recording app of choice. A small windscreen is included in order to cut down on both wind noise and percussive vocal elements.

The smartLav+ also features:

>> A Kevlar® reinforced cable, protecting your mic connection from unexpected stretching or fraying

>> High-quality omni-directional condenser mic design

>> A body design no larger than 4.5mm

TECHNICAL STUFF

Many modern smartphones are doing away with headphone jacks. Røde has developed a *VideoMic Me-L* model, powered by the iPhone's power outlet; but for non-Apple smartphones, adapters may be needed to make these suggested accessories operate properly. Before purchasing any audio recording gear for your smartphones, confirm how the accessory connects, just to assure you are picking up the right one for you.

Podcasting with Portable Recorders

Up to this point we have gone from podcasting with laptop computers to podcasting with mobile devices. Now we take everything we need to record and reduce it to a fully contained recording studio that ts comfortably in the palm of your hand.

Zoom-Zoom-Zoom: The Handy Recorder line

In 2006, Zoom Technologies (https://www.zoom-na.com/) introduced its own series of recorders that raised the bar for not just podcasting but for digital recorders across the market. With each new model, Zoom upped its own game and now offers an entire line of lightweight, unobtrusive, all-in-one solutions for portable podcasting. Welcome to the new standard.

Zoom H1n

The H1n Handy Recorder is the smallest, sleekest of the Handy Recorder series. With the *APH-1n Accessories* pack, the recorder can either attach itself to a standard mic stand or to other recording devices like DSLR cameras. Along with ease of use, the H1n offers:

>> Onboard X/Y microphones configuration

>> Recording formats include both WAV and MP3 in different bitrates and varying quality

>> Reference speaker built-in

>> 1/8-inch external mic input and 1/8-inch stereo line output

>> Records directly to MicroSD and SDHC cards up to 32GB capacity

Zoom H2n

The H2n is the next step up that may surprise you in what it can do. In the palm of your hand, with the H2n you have a surround sound recorder, a spatial audio recorder for *virtual reality* (VR) projects, and a microphone that can be used as your desktop computer's audio interface. The H2n makes portable recording a piece of cake with:

>> Five built-in mic capsules offering multiple recording modes, including Mid-Side (MS) stereo, 90° X/Y stereo, and both 2-channel and 4-channel surround

>> Recording formats include both WAV and MP3 in different bitrates and varying quality

>> Built-in studio-grade emulators including Low-cut Filter, Compressor/Limiter, Auto Gain, Tuner, Normalize, and Surround Mixer

>> 1/8-inch external mic input and 1/8-inch stereo line output

>> Records directly to MicroSD and SDHC cards up to 32GB capacity

>> 20 hours of operation on two standard AA batteries

Zoom H4n Pro

The H4n Pro sets a new standard for portable podcasting as it delivers a wide array of features building on the previous models. With the H4n Pro, you get:

>> Four-channel audio recording, able to record either in stereo or mono

>> Two XLR/TRS combo connections with both 24V and 48V phantom power options

>> Two-Input/Two-Output USB audio interface

>> Built-in studio-grade instrument effects and emulators

>> Onboard X/Y microphones able to emulate a variety of condenser and dynamic microphones

>> Capable of recording up to 140 dB SPL

>> Recording formats include both WAV and MP3 in different bitrates and varying quality

>> Records directly to SD and SDHC cards up to 32GB capacity

Zoom H5 and H6

Maybe you need more than the H4n Pro offers, and if that is the case you need to look at Zoom's H5 and H6 (shown in Figure 4-5). Both these portable recorders offer all the functionality and features found in the previous models (as you would expect), but the H5 and H6 offers more options for your on-the-go studio.

>> Interchangeable input capsules that can be swapped out as easily as the lens of a camera

>> Multichannel and stereo USB audio interface for PC/Mac/iPad

>> Recording formats include both WAV and MP3 in different bitrates and varying quality

>> Four-channel (H5) and six-channel (H6) audio recording

>> Two (H5) and four (H6) XLR connection ports with various phantom power options

>> Records directly to SD and SDHC cards up to 32GB capacity (H5) and SD, SDHC and SDXC cards up to 128GB (H6)

>> Mountable to DSLR or camcorder with optional HS-01 Hot Shoe Mount adapter (H5)

FIGURE 4-5:
The Zoom H6 is a versatile, portable recorder that raises the standard for recording on-the-go with all its incredible options for podcasters.

Portable recorders, over years of development, have established themselves as reliable, durable, and affordable options for field and on-location recording. Depending on what you need for your podcast, you no longer have to worry about remaining tethered to a studio. Portable recorders can make your podcasts happen just about anywhere!

IT'S A PREAMP! IT'S A RECORDER! KIDS. . .IT'S A PREAMP *AND* A RECORDER!

The preamp is a terriffic way to get studio-quality sound while recording portably. If you want to be able to work on a laptop and still have the ability to take your studio completely on the go, the Zoom Handy Recorders have got you covered.

The H4n Pro-H5-H6 models can all operate in the field as recording devices, and you can easily plug in an XLR microphone and conduct the interview right there on the spot. However, if you have an on-location spot secured and wish to set up more microphones for a round-table discussion, these Zoom recorders can easily plug in via an available USB port and serve as a preamp for your laptop.

And if that isn't versatile enough, you can attach any of the Handy Recorders to a USB port and use it as a dedicated microphone.

TIP

Before recording, always check that you have enough storage on your recording device. Most portable recorders will have an indicator to show how many hours or minutes can fit on the remaining storage. For mobile phones, go to your system settings and ensure you have approximately 1MB available for each minute of audio you want to record (much more for video). Chapter 9 gets into more nitty-gritty about settings and storage considerations.

WARNING

As noted earlier, RFI can be a real downer when recording. Even with portable recorders, ensure smartphones are a safe distance from your recording device. Radio, data, and cellular signals can leave a tell-tale buzzing/beeping noise and ruin your recording. Just ask Ben and Keith of *The Two Gay Geeks Podcast* (http://tggeeks.com) when they recorded a 90-minute interview with Kevin Schindler at the Lowell Observatory. Keith used his phone to prop up his recorder, with the end result sounding like a video game was being played the entire time. Fortunately, Chuck was there and had a second recording of the same interview which was unaffected. He happily shared it because that's what podcasters do!

From Cloud to Computer: Portable Audio Workflow

Now that you know how portable you want your podcast to be, how exactly will the workflow differ from the usual editing work in a studio? There are a few different approaches to consider when working portably. If you are concerned about the learning curve here, don't be. We got you, fam. It's not a dramatic switch in approaches, but more of the most efficient approach to getting audio from your mobile devices to your studio's DAW.

Getting audio from your portable recorder

After you finish recording with your portable recorder, you have audio sitting on your recorder's internal memory card. How do you get it from the recorder to the computer where you are editing your podcast?

The hard part is already done. You got great audio for your podcast, and make no debate over it — that is the hard part. Getting the audio to the computer? In the words of brave, brave, brave, brave Sir Robin at the Bridge of Death: *"That's easy!"*

1. Connect the portable recorder to an available USB port.

Recorders should have a USB cable included that physically attaches the device to your computer, as seen in Figure 4-6. If you lack such a connecting cable, your local electronics store should have what you need.

TECHNICAL STUFF

Recent computers are transitioning from USB to *USB-C* ports. Adapters are available that can quicky convert standard USB plugs to connect with USB-C ports, or provide additional ports for other devices like HDMI and native USB-C devices.

The portable recorder's interface should give you either Storage or Audio I/F as an option.

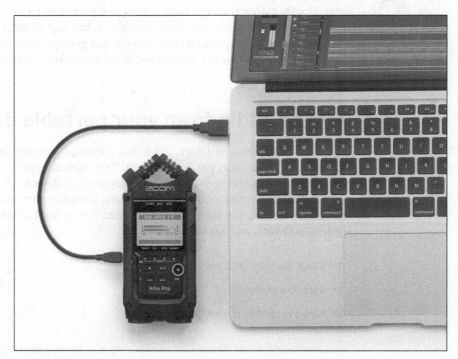

FIGURE 4-6:
Portable recorders, when connected to your computer via USB, appear on your Desktop as an external drive.

2. Select Audio Storage from the menu.

Audio I/F is the option for using your portable recorder as a USB microphone or preamp for your computer. In Storage mode, the portable recorder (and the SD card inside it) mounts onto your desktop as an external drive.

3. **Once your portable player mounts as a drive, select your new drive and then select the folder labeled in the mode you used for recording.**

Depending on the model of portable recorders, you may see folders labeled with offered recording modes. That should be where your audio is stored.

4. **Find your latest recording, and then drag it to the location where you are storing your audio sessions.**

Whether you are working on an internal drive, an external drive, or cloud storage, you now have copied your audio file source from the recorder to your workspace. You can now begin the post-production process.

TIP

It is typically quicker to transfer files from your portable recorder to your computer directly rather than use USB. If your computer has an SD card reader, you can remove the SD card from your portable record and place it in the SD card reader for faster copying. Just be sure you observe good practice and properly eject the media before pulling the SD card out of the computer and returning it to your portable recorder!

Getting audio from your portable device

With your smartphone or tablet, it's a little different environment. Instead of a hard drive or an SD card, you have internal flash memory that will quickly fill up if you are saving audio or video files directly on your device. To work with your audio recorded on your mobile device, you need to transfer it to some sort of cloud storage service. The workflow we have created here is built around the following services, software, and hardware:

>> **Cloud Service:** Dropbox

>> **Portable device:** iPad

>> **Audio recorder:** GarageBand for iOS

If you are using other devices or services, it should be easy to incorporate your own setup and adapt it for this easy-to-follow workflow:

1. **Before transferring audio, check to make sure that your cloud service's app is loaded and synced with your mobile device.**

2. **Launch GarageBand for iOS.**

If you have been working on a project, GarageBand for iOS opens on the last project you were working on or defaults to an audio recording interface.

3. **If you are in a project, tap on My Songs in the top left corner of the app.**

4. **Tap the Select option in the top right corner of the app.**

5. **Find the project you want to export and tap it once to select it.**

   The project(s) you want to export is highlighted in blue.

6. **Tap the Share option in the top left corner of the app.**

   The Share function (shown in Figure 4-7) accesses which apps are able to share the media you are about to export. If you cannot find the app you want to share to, find the More option to add your cloud service app.

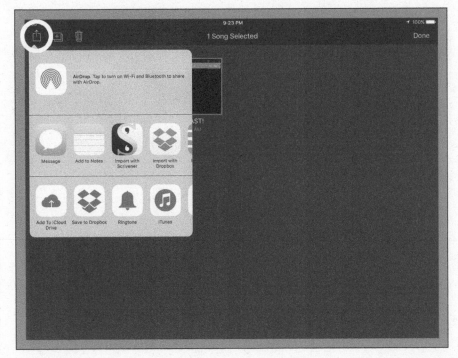

FIGURE 4-7:
From the Share icon (circled in the top left), apps are offered where your media can be shared.

7. **Tap your cloud service app.**

   In the case of Dropbox and iPad, tap the Save to Dropbox option to begin the export process.

8. **Edit the Info for the file and then select Audio Quality from the offered options. Tap Share to begin the exporting process.**

   In GarageBand, you can select from four different MP3 formats, Apple Lossless (m4a), and AIFF.

9. **Select the location where you want to save your media and then tap the Save option in the top right corner of the Save window.**

Your media has now been exported on to your cloud service and is waiting for you to edit or prepare for uploading.

WARNING

While you can record and export audio to MP3, it is never a good idea to make an MP3 from another MP3. It is the audio equivalent of making a compressed JPEG image from another JPEG image. Always strive to record in a raw, uncompressed format such as AIFF or WAV files. It takes more space initially but produces better results.

2

The Hills Are Alive with the Sound of Podcasting

Understand the steps involved before hitting the record button, such as topic, posting frequency, and episode duration.

Unlock the art of the interview, including interview prep, what questions to ask (and not to ask), what software options are available for interviewers, and how to query for interviews.

Record with confidence after sound checks and level checks and ensure that your software is ready to record.

Add production value to your podcast with intros, outros, bed music, and simple editing.

IN THIS CHAPTER

» **Finding a voice for your podcast**

» **Using an outline or script**

» **Deciding your podcast length and schedule**

» **Understanding legal issues**

Chapter **5**

Before You Hit the Record Button

Tune to a classical radio station (and when we refer to "classics" here, we mean Beethoven and Haydn, not the Beatles and Hendrix) and listen to the DJs — oh, sorry, the *on-air personalities* — featured there. You'll notice that they're all speaking slowly and articulately, mellowed and obviously relaxed by the melodic creations of greats such as Mozart, Wagner, and Joel. (Yes, Billy Joel has a classical album — a pretty good one, too!) Although the on-air personalities of your local classical music station all sound alike, they sound dramatically different from the wacky Morning Zoo guys on your contemporary hits radio station who sound as if they're on their eighth cup of espresso.

When you hear people talk about *finding your voice* in broadcasting, that's what they mean. You come to an understanding of what your average audience wants (and to some degree, expects), and then you meet that need. This chapter helps you develop the voice and personality you want to convey when podcasting.

After you discover your voice, you will want to get ready for the show. This chapter shows you what to do to prepare for smooth and easy podcasts that (one can hope) will be glitch-free during the recording process. Preshow prep is not only important, but also essential in making a feed worth catching. Even the most spontaneous of podcasts follows a logical progression and general direction, remaining focused on the podcast's intent.

Choosing a Unique Topic for Your Podcast

The first thing to understand about podcasts is that it isn't all about being "number one" in your chosen podcast genre. Sure, some podcasts do vie for top honors on various polls, but instead of worrying about analytics and accolades (covered in Chapters 12 and 13), think about what will make your podcast uniquely worth your effort and your listeners' time. The point in launching a podcast isn't always "I want to do something totally new..." but more about "What do I have to say about this topic?"

As of January 2020, according to Oberlo (`https://www.oberlo.com/blog/podcast-statistics`), there are currently 850,000 active podcasts — and that number keeps going up. You're going to need something unique to stand out in the crowd.

Here are some ways you can create a unique podcast:

>> **Identify your audience.** Understand who your audience is and what their interests or issues are. You can do this by talking to like-minded people online and in real life. Consider participating in a Facebook group or other online community related to your topic. Remember, it's all about the audience. Once you understand them, you'll be a lot better off.

>> **Study other podcasts.** Before you can figure out what will make your podcast unique, check out other podcasts related to your own. The best way to find out what makes a podcast worthwhile is to subscribe to a few feeds that pique your curiosity.

Listen to these feeds for a few weeks and jot down what you like (and don't like) about them. From the notes you take, you might find your angle. Keep in mind that downloading and listening to other podcasts should be educational and constructive, not a raid for fodder on your own show.

WARNING

Don't steal content, special effects, or unique segments (like "On This Day in Tech History" or "Writers Off the Clock") from another podcast. Approach others' podcasts as you would someone's online content. It's okay to be inspired, as long as you don't make your podcast a carbon copy of your inspiration's work. Make it your own! When you have your podcast up, avoid criticizing another podcast in your own; criticizing someone else's work is no way to better yours. Stay on the pod-sitive side.

>> **Pick a topic you know.** Whether you've decided to take on the topic of music, religion, or technology, the best way to make your podcast unique is to find an angle you're comfortable with (Polka: The Misunderstood Music, Great Travesties of Sports History, Forgotten Greats of Science Fiction). There's also

the possibility that your initial show may inspire an additional angle so unique that you'll have to start another podcast specifically to address that audience.

>> **Speak confidently.** Don't apologize for being "yet another podcast on. . ." or point out what you are doing "wrong" compared to others. What makes a podcast fun is the passion and the confidence you exude when the mics are hot. Address your topic with authority and energy, and enjoy your time recording. If you have a blast making a podcast, your audience will enjoy it along with you. (That confidence might even inspire others to create podcast themselves.)

Figure 5-1 is a good example of a podcast that focuses on what it knows.

REMEMBER

The content you bring — regardless of what genre it's in — is unique because it is your podcast. It's your voice, your angle, and your approach to whatever intent you pursue. Provided you maintain a high confidence level and genuinely enjoy what you're doing, people will tune in and talk to other listeners about what you're podcasting.

Finding Your Voice

The broadcasting industry might not want to admit to this, but podcasting and commercial radio share a lot in common. In the early days of what is now a major radio genre, talk shows were reserved for National Public Radio and news stations. In general, they were pretty dry and lackluster, bringing their listeners the news, weather, and daily topics that affected the world — but nothing particularly unusual or exciting.

Then a guy named Howard Stern came along and changed everything in this once-tiny niche! You can love him, you can hate him — you can claim to hate him when secretly you love him — but Stern completely turned around what was considered AM-only programming. Now talk radio is big business. Some personalities are just out to entertain, other hosts deep dive into lifestyles and subjects of interest, while others use it to voice their political viewpoints.

A majority of podcasting is just that: talk radio. Each podcast has a different personality and appeals to a different market. Finding your voice is one of the most challenging obstacles that you (as a once-and-future podcaster) must clear. Even if your podcast's aim is entertainment, you have a message you want to convey. That message will influence the voice you adopt for your podcast. If you're podcasting an audio blog about life, its challenges, and the ups and downs that you encounter, then maybe a soft tone — relaxed and somewhat pensive — would be appropriate. But if you decide to go political — say you're the Angry Young Man who's fed up with the current business on Capitol Hill — then it's time to fine-tune the edge in your voice. That's what you need for a podcast of this nature.

After you discover the passion your podcast is centered around (see the preceding section for tips on how to do that), here are some ways to *find your voice:*

>> **Record your voice and then listen to what it sounds like.** It astounds us how many people hate listening to their recorded voice. It's a fear akin to getting up in front of people and speaking. When finding your voice, though, you need to hear what your current voice sounds like. Write a paragraph on your show's subject. Then read it aloud a few times and find a rhythm in your words. Expect the following:

- Talking too fast

- Swallowing small, one-syllable words like *to, in*

- "Pause words" or "stop words" like *um, ah, y'know,* and *right*

- Ignoring commas, thereby creating one long, run-on thought

- Lip-smacking, heavy breathing, and the unavoidable *ahs* and *ums*

You can edit out some of these problems (see Chapter 8), but you should grow accustomed to hearing your own voice because you'll hear yourself again and again . . . and again . . . during the editing process. The more familiar you are with how your voice sounds, the easier time you'll have editing your podcasts before publishing them online.

>> **Play around with the rhythm of your speech.** You don't have to be an actor to podcast, but you can apply some basics of acting when you're recording. One of these basics, as Tee's acting mentor Glyn Jones told him, is to "Make a meal of your words." This means to play around with the rhythm of your speech. When you want to make a point, slow down. If you're feeling a tad smarmy, pick up the pace. Above all, be relaxed and make sure you don't sound too contrived or melodramatic.

>> **Speak clearly.** Another simple trick from the acting world to add to your arsenal is to open your mouth wider. Many people talk with their mouths mostly closed, but by opening up your mouth, you can gain clarity. So, when making a meal of your words, it is good manners to chew with your mouth open.

>> **Speak with confidence.** Yes, we're saying this again, because it bears repeating: Speak confidently about your topic. No one is going to believe in what you have to say if you don't believe in yourself. It may take a few podcasts to find a groove, or you might hit the ground running and have a podcast that immediately takes off. Just speak with conviction and allow yourself to shine.

>> **Develop your podcasting personality.** After you know what you sound like when you record, here's where you develop your podcasting personality. Is your persona going to be light, fun, and informal, or something a little edgy, jaded? Is your message taking an angle of marketing, politics, or religion? Or are you podcasting a love of music, science, or your Macintosh? Your persona should generally match the theme of your show. If you're doing a show on classical music, a persona of a morning radio DJ probably isn't going to work. If you're taking a light-hearted look at politics, you may want to have a little more levity in your tone and pace than a funeral director.

What if I hear more than one voice?

One of our favorite ways to podcast is with guests in-studio or cohosts where more than one podcaster gets on mic. While there's something to be said for the single voice doing a monologue or perhaps doing interviews, the show dynamics change quite a bit when you get multiple people gathered together over your favorite topic. For one thing, it's a lot easier to carry on a conversation! Another bonus is with the right dynamic between hosts, an energy is created that sub-scribers see and hear in every episode. Along with the guidelines described in this and other chapters, there are some specific things to be aware of when doing a show with multiple guests.

>> **Have a mixer with enough channels.** Back in Chapter 2, we talk about gear. Remember the mixer? The mixer becomes a crucial piece of equipment when

cohosts become part of the production. You can try the one mic, two voices approach, but the end result is hard to control and mix in post-production. For the best sound and optimal control, each person needs their own microphone. This means XLR connections, not USB, for microphones. Two hosts, and you'll need two channels. Four hosts, four channels. See how that goes? And don't forget, you may want a few extra inputs for music, sound effects, and more.

>> **Make sure everyone can hear.** You're wearing headphones when you record. So should your guests, especially if drop-in's are included in your recording. It's not only fair, it's practical that everyone hears the same thing. Each guest needs their own set of headphones. Before you run out and get a cheap "Y" cable to split the signal, realize that with each split, the audio signal degrades. To keep the investment economical, invest in a *stereo headphone amplifier* for about $25 that takes the headphone signal and splits (while boosting) it in to four separate channels. Then pick up from BSW a 5-pack of Audio-Technica headphones (`https://bit.ly/AT5pack`) for you and your cohosts or guests. You'll find this investment will serve you and your podcast well.

>> **Always do your prep work.** Even after "centuries" of podcasting, there are still gremlins in our audio systems. We can record on Saturday afternoon and come back Sunday night only to find audio levels have been adjusted. Okay, it could be the cats playing with the mixer settings in the middle of the night, but it never hurts to check your audio (and video) settings before each recording.

>> **Have one director.** This is the person in charge of your show's flow, timing, and in some cases coming up with clever segues to jump from one topic to the next. Usually this is the person at the mixer, but not always. It may even be someone off mic (or camera) giving hand signals. In some cases, this may be a baton passed from person to person in the cast. You'll find what works best for your group. The podcaster calling the plays serves as a moderator. It is your job to keep the energy up, the conversation going, and keep the episode on track.

>> **Give everyone some air time.** Similar to the previous item, the director may also need to make sure everyone gives everyone else a chance to talk. Different people bring different things to your show. Some people may be passionate and outspoken (and some may be considered an unstoppable train) while others don't want to interrupt so they wait their turn. Encourage your guests to play fair and give everyone a share of the air time. We recommend discussing this among your cohosts before it becomes a problem.

>> **When a guest is in-studio or on the line, give the guest(s) the majority of air time on that episode.** Both Chuck and Tee have seen and heard their fair share of interviews gone bad. It can be something as horrific as the host or hosts not knowing (or caring) to do any research on the guest. When a guest is

on the docket, remember that the episode is no longer yours. It's theirs. For more on interview techniques, take a look at Chapter 6, coming soon.

>> **Provide guest orientation.** Before recording, let guests know the format and flow of your show. You don't need to go in to all the details, but give them a general idea when they can start talking, how long the show runs, and if you want them to hang around after thanking them on the recording.

>> **Make sure everyone can see everyone else.** It's been said that as much as 93 percent of our communications is nonverbal. Even if you are doing an audio podcast, you want to be able to see each other during the conversation. As seen in the seating configuration depicted in Figure 5-2, being able to read each other's nonverbal cues makes it easier to get talk time in a conversation — well, most of the time. If you are not in-person, use Google Hangouts, Discord, Zoom, or some other video conferencing to see each other — yes, even for audio podcasts!

FIGURE 5-2:
Sightlines matter when you have cohosts or guests in-studio. Configure your studio to make sure everyone can see one another in order for eye contact, silent signals to pick up the pace or slow down, or let the director know they have something to say.

REMEMBER

Be aware that your show will be longer as you include more guests in the conversation. If you want to keep your show length consistent — a good recommendation in our book — then include fewer topics than you expect. Part of Chuck's Saturday ritual for building out the Sunday night episode is to find out how many people are coming over so he can add or remove topics accordingly. For Tee, it's coordinating with his cohosts who is wanting to take the lead on a specific topic. Breaks offer a moment for anyone to tell him "Mind if I take the lead in the second half?" If you really don't care about length, then just realize that more guests will make for a longer show and plan accordingly. Communication is essential with cohosts and in-studio guests.

Deciding Whether You Need an Outline or Script

What method works best for you? A full script and hours of prep time, or a single note card and two clicks of the mouse — one for *Record* and another for *Stop?* Both approaches work, depending on the podcaster's personality. It could be said that there's little difference between a writer and a podcaster: Some writers prefer to use an outline when putting together a short story or novel; others merely take an idea, a few points, and a direction, and then let their fingers work across the keyboard.

If you decide to work with a script, it's a good idea to invest some time into *pre-show prep*, simple preparation for what you're going to say *and* how you're going to deliver it. Depending on your podcast, though, prep time may vary. Here are a couple of examples of how dramatically different prep time can be for different podcasting situations:

>> For their podcast *The Brit and Yankee Pubcast,* Phil Clark and his crew do very little prep — usually just enough to get some basic facts about the drink of choice for that show and perhaps set up a location and interview with the brewmaster. After he does minimal orientation with the guest panel, he's ready to record. You really need to know your subject and have good chemistry with your show participants to make a minimal plan like this turn in to a good show, but it can work.

>> On the other side of the spectrum is one of the original podcasts, *The Radio Adventures of Doctor Floyd,* a 10-minute show in the style of old-time radio with a modern, educational, comedic spin. Grant Baciocco and Doug Price (see Figure 5-3), from 2004–2010 and across eight seasons, have every show carefully scripted. Depending on the historical research required, Grant would spend anywhere from 1 to 3 hours doing preshow prep. The careful scripting comes in real handy when "Doctor Floyd" has celebrity actors playing a part in the show.

FIGURE 5-3:
The Radio Adventures of Doctor Floyd's Grant Baciocco (left) and Doug Price (right) take their comedy seriously, and that means plenty of preshow prep!

Preshow prep can range anywhere from jotting a few notes on a napkin to writing a complete script with full sound effects — regardless of show length. So how far should your prep go technically? That depends on what your podcast needs. Outlines and scripts will keep you on track with what you want to say, serving as roadmaps you use to keep moving smoothly from Point A to Point B.

Whether you're a napkin scribbler, a script writer, or somewhere in between, if you've never done any kind of planning like this, the secret to efficient preshow prep can be boiled down to three disciplines:

» **Habit:** Many podcasters, especially podcasters emerging from corporate offices, prepare for podcasts in the same manner as business presentations. They jot down essential points on note cards to keep the podcast on track, but the points are the only material they write beforehand. You can easily apply your organizational skills from the workplace to the podosphere.

» **Talent:** Some podcasters are truly the Evel Knievels of podcasting, firing up their mics and recording in one take. These podcasters tend to have backgrounds in live entertainment, deciding in a moment's time when a change of delivery is required. This is a talent of quick thinking, and although it keeps material spontaneous and fresh, it's a talent that must be developed with time.

>> **Passion:** Passion is a driving force with a majority of podcasters that keeps their podcasts spur-of-the-moment. With enough drive, inspiration, and confidence in their message, they keep their prep time to a minimum because podcasting isn't a chore but a form of recreation.

Determining a Length for Your Show

If you've been using this chapter to develop your podcast, you've made serious progress by this point in getting your preshow prep done. Now you're ready to podcast, right?

Well, no. Have you thought about how long your show's episode is going to run? No? Okay then, check out the following sections.

TIP

You may hear some veteran podcasters say "Your podcast should never be longer than 30 minutes because that is the time of an average commute." Chuck has heard actual listener feedback asking "Can you make your show the length of my commute?" Oh boy! Tee's response to this has not changed since 2005: "Obviously, you don't live in the Washington, D.C. Metro area, do you?" Yes, maybe the average commute from sea to shining sea is 30 minutes, but that should not be a set-in-stone template. There are shows where each episode is 60 seconds, others where the host(s) decide they need to go on for hours (thank you to whomever invented the pause button!), and still others where each episode length is variable. The bottom line — you decide how long it takes you to podcast, and remember, there are going to be those "super-sized" episodes that occasionally come along.

The hidden value of the short podcast

There are many podcasts that run under 10 minutes where hosts deliver their message and then sign off only moments after you thought they signed on. While on average — and this is more like an understood average as noted above, not really a scientific, detailed study of all the podcasts out there — a podcast runs from 20 to 30 minutes per episode. So, what about these 10-minute vignettes? Does size matter? Does time matter? (Woah. Deep.) Is there such a thing as too short a podcast?

Here are some advantages in offering a short podcast:

>> **Shorter production time:** Production time may be reduced from a weeklong project to a single afternoon of planning, talking, editing, and mixing. With a quick and simplified production schedule, delivering a podcast on a regular basis — say, every two weeks, weekly, or twice a week — is easier.

- » **Faster downloads:** You can be assured — no matter what specs you compress your audio file down to — that your podcast subscribers will always have fast and efficient downloads.

- » **Easy to stay on target:** If you limit yourself to a running time of less than 10 minutes, you force yourself to stick to the intent (and the immediate message) of your podcast. There's no room for in-depth chat, spontaneous banter, or tangents to explore. You hit the red button and remain on target from beginning to end, keeping your podcast strictly focused on the facts. Shakespeare said, "Brevity is the soul of wit." Considering his words, ol' Bill would probably have podcast under 15 minutes if he were alive today.

Nothing's wrong with keeping a podcast short and sweet. In fact, you might gain more subscribers who appreciate your efficiency.

A little length won't kill you

Now with that quote from the Bard about brevity, you might think, "Shakespeare said *that?!* Before or after he wrote *Hamlet?*" That's a good point because Shakespeare did have a number of his characters say, "My lord, I will be brief . . ." and then launch into a three-to-four-page monologue.

So what if Shakespeare decided to be brief in his podcast? Would he get any subscribers if his show ran longer than half an hour? What if he broke the 60-minute ceiling? Would the Podcast Police shut down his show?

Listeners, on reading your show notes and descriptions, should be able to figure out the average running time of your show. On a particular topic, some podcasts can easily fill two or even three hours. Huge productions have some definite advantages:

- » **If the show is an interview, you have anywhere from two to three hours with an authority.** It's something like having a one-on-one session stored on your computer or MP3 player. From shows like SyFy Wire's *The Churn* (see the upcoming sidebar), if a guest is part of the podcast, you can rest assured your podcast will go a little longer than 30 minutes — and sometimes it should.

  WARNING

  Be careful with this one. Shows and interviews that ramble aimlessly run the risk of losing audience attention. We talk more about good interview practices in Chapter 6.

- » **You're allowed verbal breathing room.** Discussion stretching past the 30-minute mark allows you and your cohosts or guests to break off into loosely related banter, widening your podcast's focus and sparking discussion that can lead in other directions.

FREE FALLING INTO *THE EXPANSE*

SyFy, since the early days of podcasting, has been employing the medium as an "enhanced experience" to its leading shows. In the first and second editions of *Podcasting For Dummies*, *Battlestar Galactica*'s Executive Producer Ronald D. Moore and SyFy (then called SciFi) were featured for hosting hour-long episodes, similar to director commentaries, providing an inside look at what went into the episode airing that day. The companion podcast of *Battlestar Galactica* was so successful that SciFi broadened its scope to include scriptwriting sessions and guest appearances from cast members.

Since then, SyFy continues to podcast as a means to bring fans deeper into their most popular shows, and *The Churn* is such a podcast. *The Churn*, hosted by Ana Marie Cox and Dan Drezner, takes a deep dive into *The Expanse,* a series that started on SyFy and continues on Amazon Prime. Along with getting into the heads of the series creators, *The Churn* also invites actors, visual artists, and special science guests like *Bad Astronomy*'s Phil Plait to talk about issues addressed in the show from accuracy of physics to theoretical politics stretching across the solar system. If you are a fan of this incredible science fiction series, *The Churn* is a must-have companion for watching *The Expanse*.

REMEMBER

The cost of podcasts longer than 30 minutes is in bandwidth and file-storage — issues that smaller podcasts rarely, if ever, have to deal with. See Chapter 10 for a discussion of the bandwidth demands on your server.

Finding that happy medium

Is there such a thing as middle ground in the length of a podcast? How can you find a happy medium if podcasters can't agree on a standard running time?

The happy medium for your podcast should be a sense of *expectancy* or *consistency*. For example, in Tee's award-winning podcast of *Tales from the Archives*, the running times for each episode are across the board — the shortest clocking in at just over 30 minutes, and the longest weighing in at over an hour. His audience, however, understands this is a *podiobook*, an audiobook presented in a serialized format. Readers understand that chapters and short stories vary in size, so it's no surprise when a podiobook follows suit. Some of the episodes are short and sweet while others push the length limits expected from literature.

TIP

Podiobooks aren't the only genre that variable length works well for. If your podcast deals in do-it-yourself home improvement, explaining the construction of a bookshelf will be a far shorter show than one about adding an extension to your deck.

Give yourself some time to develop your show, your voice, and your direction. If you build some consistency and expectation for your audience, it's easier to introduce a little variation or even a happy medium into your running time.

Mark Your Calendar: Posting Schedule

You've got a format for your show. You've got an idea about its running time. Now you have to figure out when your show is going to go live. What is the best pace to set for your podcast? What are the advantages to posting frequently versus posting on occasion? How often will you be dropping your podcast into your feed?

There are four different kinds of posting schedules, some easier to maintain than others. Your podcast, depending on the planning and running time you set, will dictate how often you post. There is no "sure schedule" to podcasts. What matters is setting a schedule and maintaining it.

TIP

A good way to know what posting schedule works best for you is to sit down and brainstorm on show topics. If you rattle off several ideas, rapid-fire, you may be looking at a frequent schedule. If your ideas are reliant more on current events and their outcomes, you may space out your episodes. See how quickly you can come up with ideas, and from there, make a decision on a reasonable posting schedule.

Posting daily

The demands of podcasting can be daunting between recording, editing, and posting. There are those who have figured out a way to minimize production, whether it is keeping the recordings raw and unedited or employing a studio with a crew or something in-between and posting on a daily schedule. Every day, a new episode appears in your podcast app. Every day. That's a lot of content to sift through if you subscribe to a long-running podcast. Daily podcasts, though, do not necessarily follow a linear path. In other words, you do not have to go back to Episode 1 to understand the flow and the atmosphere of a podcast. Just jump on in and enjoy!

If you think podcasting is daunting, podcasters like Nathan Lowell on *Today on My Morning Walk* (http://www.nathanlowell.com/tommw/), the high-energy *Geek Radio Daily* (http://geekradiodaily.com), and *The Washington Post*-powered *The Daily 202's The Big Idea* (https://www.washingtonpost.com/podcasts/daily-202-big-idea/) undertake the challenge and produce new content every day. A payoff to answering this challenge is building an audience. With so much content to share, your community should grow quickly.

WARNING

Regardless of your intended schedule, life sometimes deals us an unintended hand. Just let your audience know. They'll understand. We've found that doing this can build loyalty. Let's face it, the majority of podcasters are doing this as a "second job." We have family that needs attention, day jobs that require business trips, and a host of other things that might crop up, scheduled or otherwise. You don't have to reveal personal details if it's not appropriate; however, a quick message to your audience is always polite.

Posting weekly

Perhaps the most common of schedules for podcasts is weekly posting, like you would with a popular television show. Perhaps not as demanding as the daily schedule, this schedule means a commitment to producing new content at least once a week. You will want to make certain the content is there before you launch and find workflows that make your production schedule more efficient. You can do this by maintaining a buffer of content, seen often in gaming podcasts like *Steam Rollers Adventure Podcast* (http://riggstories.com/srap/) and *So Many Levels* (http://christianaellis.com/so-many-levels-a-dd-podcast/) that record gaming sessions that can last for an hour (or longer) and then present them in a serialized format. Pick a day of the week and make that your day. That will be the time your listeners or viewers will be expecting your next episode.

Other podcasts, like George Hrab's *The Geologic Podcast* (http://www.geologic podcast.com), shown in Figure 5-4, and *Grammar Girl's Quick and Dirty Tips* (http://www.quickanddirtytips.com/grammar-girl), seem to just happen spontaneously. There's some planning that goes into these podcasts, but talent also comes to play. If you can get behind the mic and feel right at home, you can keep up with the weekly schedule.

Posting biweekly (or fortnightly for our friends in the Commonwealth)

Say you have a reasonable amount of content, but not a whole lot of free time. Or maybe you want to podcast but are concerned that you will burn through the content before you can come up with new ideas and directions for your show. A biweekly posting schedule provides you with a comfortable alternative to the regular demands of a weekly production. This may mean the audience response and the timeliness of your podcasts may be lacking when compared to a more frequent posting schedule, but this schedule is easier to maintain in case a weekly schedule is difficult for you to maintain. Podcasts like *The Topic is Trek* (http://www.the topicistrek.com) find a good life-work-podcast balance with the biweekly schedule.

FIGURE 5-4:
The Geologic Podcast, hosted by musician, voice actor, and author George Hrab, has stuck with a weekly schedule since its launch. His endurance and creativity have spawned hundreds of shows of comedy, music, and critical thinking.

TIP

Whether working daily, weekly, or biweekly, you might find life stepping in the way of your production schedule. This is why having a concept of seasons should be considered for your podcast. In the case of Tee's steampunk podcast, *Tales from the Archives,* seasons are defined by 10 to 12 short stories. If you see a break coming in your production, make sure to let your audience know.

Posting monthly

What are the benefits of a monthly podcast, aside from the relaxed production schedule? It's easy to see one of the challenges in working on a monthly schedule: nurturing the audience. Posting only monthly makes it difficult — not impossible, but difficult — to build a community over content that only happens once a month. Additionally, the timeliness of a monthly podcast is almost nonexistent as news headlines happening weeks ahead of a recording session is impossible to comment on in a timely fashion.

What does make a monthly podcast schedule appealing? Longer-than-usual running times on topics inspired by the recent weeks' headlines. Such is the case with shows like *The Monthly Reset* (https://anchor.fm/monthly-reset/), *The Monthly Movie Show* (https://bit.ly/monthlymovie), and *Art Monthly Talk Show* (https://www.artmonthly.co.uk) where the hosts enjoy the conversation, tangents and all, recording episodes where an hour or less would be considered a "short" episode. As monthly podcasts have so much time to prep, hosts can gather plenty of resources and backstories as well as their thoughts. So if you have a subject that needs time for research and discussion, posting monthly may be a better option for you.

TIP

If you find yourself with an intermittent schedule, avoid apologizing for the absence. Over the years, both Tee and Chuck have listened to a number of podcasts that have somewhat random schedules. When we hear from those wayward podcasters, we're excited to hear from them again. It's a bit of a letdown when the first words we hear are "Sorry for not producing a show as much as I would like." Instead, consider telling us how glad you are to be back or just don't mention it at all. Get right in and start your delivery as if nothing happened. We forgive you!

I Hear Music (and It Sounds Like Police Sirens!)

In creating your own podcast, something that will give your show an extra punch or just a tiny zest is the right kind of music. Although our skills and tastes range from classical to jazz to rock 'n' roll, both Chuck and Tee appreciate and understand the power of music and what it can bring to a podcast.

Your authors also understand and appreciate the law. Although you may think it's cool to "stick it to the man" and thumb your nose at Corporate America, the law is the law, and there are serious rules to follow when featuring that favorite song of yours as a theme to your podcast.

WARNING

We want to make this clear as polished crystal — we are not lawyers. We're podcasters. We've looked up the law on certain matters so we know and understand what we're talking about, but we are not lawyers. We can tell you about the law and we can give a few simple definitions of it, but we are not giving out legal advice. If you need a legal call on a matter concerning your podcast — whether it concerns the First Amendment, copyright issues, or slander — please consult a lawyer.

The powers that be

The government still regards the Internet even today as a digital Wild West, an unknown territory that's avoided regulation for many years, granting those who use it a true, self-governed entity where ideas, cultures, and concepts can be expressed without any filtering or editing, unless it comes from the users themselves.

Does this mean we podcasters are free to do as we please? Well, no, not by a long shot. There are some rules and regulations that even podcasts must follow. There are also organizations that both broadcasters and podcasters must pay attention to.

The following organizations all have influence on the destiny of podcasting, and it's only going to benefit you as a podcaster to understand how their legislation, activities, and actions are going to affect you.

The Federal Communications Commission (FCC)

The Federal Communications Commission, or FCC (https://www.fcc.gov) is the watchdog of anything and everything that gets out to the public via mass communications. The FCC keeps an eye on technology development, monopolies in the telecommunications industry, and regulating standards for telecommunications in the United States and its territories. It is most commonly known for enforcing decency laws on television and AM/FM radio.

For podcasters, the FCC can't regulate what is said (yet) because it doesn't consider the Internet a broadcasting medium. However, given existing legislation to reduce *spam* (junk email) and the ever-growing popularity of podcasting among mainstream broadcasters (such as iHeartMedia, NPR, and ESPN), it may not be long before the law catches up with technology.

The Recording Industry Association of America (RIAA)

Shawn Fanning. Does that name ring a bell? It was Fanning who lost his battle against the Recording Industry Association of America, also referred to by its more common acronym RIAA (http://www.riaa.com), when he contested that his file-sharing application, Napster, in no way infringed on copyright laws and was not promoting music piracy. The RIAA led the charge in shutting down the original Napster and continues to protect property rights of its members — as well as review new and pending laws, regulations, and policies at the state and federal level.

The RIAA will have a definite say as to why you cannot use a selected piece of music for your podcast. Simply put, it's not your music. Sure, you downloaded that album from Apple Music, but the music you listen to is under the condition that you use it for listening purposes only. (Didn't realize there were conditions involved, did you?) This means you can't use it as your own personal introduction that people will associate with you. And, no matter how appropriate your favorite song is, you cannot use it as background music. Unless you're granted licenses and you pay fees to the record labels and artists, you're in copyright violation when playing music without permission.

TIP

One way of getting music for themes, background beds, and segues is to look into what musicians and podcasters refer to as *podsafe music*. This is professionally produced music from independent artists who are offering their works for podcasting use. The demand for podsafe music has been so high that several sites like the Free Music Archive (see Figure 5-5) have been launched (`https://freemusic archive.org`), offering a wide array of genres, artists, and musical works. Today several other sources of podsafe music exist, including Digital Juice (`https://www.digitaljuice.com`), Neosounds (`https://www.neosounds.com`), Sound Stripe (`https://www.soundstripe.com`), and Instant Music Now (`https://www.instantmusicnow.com`). Find out more about podsafe music, the conditions of using it, and how it can benefit your podcast.

FIGURE 5-5:
The Free Music Archive is an interactive library of high-quality, legal audio downloads directed by WFMU.

The Electronic Frontier Foundation (EFF)

In addition to the big dogs who are passing the laws and legislations to restrain your podcasting capabilities, a group is looking out for you, the podcaster, with Science Fiction author and tech activist Cory Doctorow stepping forward as one of its more outspoken members. The Electronic Frontier Foundation, or EFF (`https://www.eff.org`), is a donor-supported organization working to protect the digital rights of the individual; to educate the media, lawmakers, and the public on how technology affects their civil liberties; and uphold said civil liberties if they're threatened.

A good example of EFF's mission is its involvement in various legal cases concerning URL domain registration and *cybersquatters* (individuals who buy desired domains and then hold on to them, waiting for the highest bidder). The EFF stands for the rights of legitimate website owners who happen to own a domain that a larger corporation would desire to use.

The EFF, provided you have a strong case to contradict the findings of the RIAA and the FCC, will stand up for you and give your voice a bit of power when you're standing up to a corporate legal machine.

Creative Commons (CC)

Founded in 2001, Creative Commons (CC) is a nonprofit corporation dedicated to helping the artist, the copyrighted material, and the individual who wants to use copyrighted material in a constructive manner but may not have the resources to buy rights from groups like the RIAA.

Copyright protection is a double-edged sword for many. On the positive side of a copyright, your work is protected so that no one can steal it for their own personal profit, or if someone makes the claim that you're ripping off their work, your copyright is proof that your egg came before their chicken. That's the whole point of the copyright — protection. The downside of this protection is that people now must go through channels for approval to feature your work in an educational or referential manner; and although you're given credit for the property featured, there's still a matter of approvals, fees for usage, and conditions that must be met. Also, many contributors just want to share their work with others on no other terms but to contribute and share with the world. Copyrights complicate this.

This complication of the digital copyright, protections, and desire to exchange original creations brought about Creative Commons (https://creativecommons.org), one of the attributions pictured in Figure 5-6. It's dedicated to drafting and implementing via the Internet licenses granting fair use of copyrighted material.

In the case of the podcaster, you want to offer your audio content to everyone, not caring whether listeners copy and distribute your MP3. As long as the listeners give you credit, that's all fine and good for you. CC can provide you with licenses that aid you in letting people know your podcast is up for grabs as long as others give credit where credit is due. CC provides these same licenses for artists and musicians who would not mind at all if you used their music for your podcast.

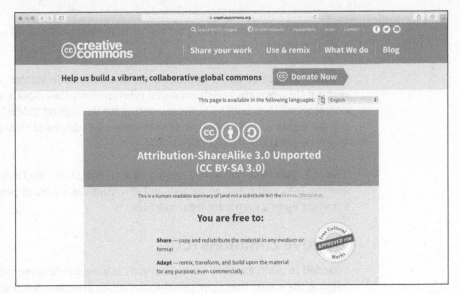

FIGURE 5-6:
Creative
Commons offers
free licenses for
use of original
content in
podcasting.

CC licenses are made up of permission fields:

>> **Attribution:** Grants permission for copying, distribution, display, and performance of the original work and derivative works inspired from it, provided credit to the artist(s) is given.

>> **Noncommercial:** Grants permission for copying, distribution, display, and performance of the original work and derivative works inspired from it for noncommercial purposes only.

>> **No Derivative Works:** Grants permission for copying, distribution, display, and performance of the original work only. No derivative works are covered in this license category.

>> **Share Alike:** Grants reproduction of the original work and also allows derivative works *if* they are also released under a similar Creative Commons license.

These four fields can be used as stand-alone licenses or can be mixed and matched to fit the needs of the podcaster or the artist offering content for the podcast.

The CC and its website give details, examples, and an FAQ page that answers questions concerning the granting of licenses for use of protected content. Just on the off-chance you don't find your answer on the website, it gives contact information for its representatives. CC is a good group to know and can open opportunities for you to present new and innovative ideas and works in your podcast.

I can name that tune . . . I wrote it!

Using almost anyone else's music for your podcast can be an open invitation for the RIAA to shut it down. This is primarily to protect the artist's rights. Think about it — how would you feel if you were producing a popular podcast, receiving praise from all over the world, and while you're thinking about ways of taking the podcast to the next level, you turn on your radio and hear your podcast being broadcast on a top-rated radio station. Soon, your podcast is all the rage on the broadcasting airwaves — and you haven't made dollar one.

The same thing can be said for artists and their music. They work hard to produce their work, and now podcasters are using their music to brand their shows, not bothering to compensate the artists for their efforts. Artists love to say that they "do what we do for the love of the craft" but in the end, it's their *work* and artists have to pay the bills, too.

So how can you use a piece of music without suffering the wrath of the RIAA or FCC? Ask permission of the artist? Only if the artist owns the rights to the music and the recordings. Otherwise, you also need to get written permission from artists, musicians, record labels, producers. . .

REMEMBER

The best way to avoid the legal hassles is to avoid copyrighted material that is not your own.

If you want to use published pieces that aren't royalty-free, ask the artist directly (if you can) for permission to use that music on a regular basis. Compensation to the artist may come in the form of a promotion at the beginning or end (or both) tags of the podcast. As long as you have written permission from the artists and the artists have the power to grant it (that is, they haven't signed the power over to their label or publisher), you should be able to use their work to brand your show or feature them on your podcast. (If you're not sure whether you have the appropriate permission, you may want to consult an attorney.) This is usually acceptable with independent artists because, in many cases, they also own the record label. Confirm this with artists. Otherwise, you run into the same legal issues if you were to use music recorded by Queen, Bruno Mars, or U2.

TIP

You can always offer your podcast as a venue for the musician to sell his or her work. Dave Slusher of *Evil Genius Chronicles* (www.evilgeniuschronicles.org) obtained written permission from the Gentle Readers to use its music as intro, exit, and background music for his podcast; in return, Dave's model has been working well for the Gentle Readers as well as artist Michelle Malone (https://www.michellemalone.com). After her music was featured on Dave's podcast, her sales spiked — both through her website and on Apple Music!

I'll take the First: Free speech versus slander

Words can (potentially, at least) get you in just as much trouble as music. The legal definition of *slander* is a verbal form of defamation or spoken words that falsely and negatively reflect on one's reputation.

So, where does podcasting fit into all this? Well, the Internet is a kind of public space. Think about it — before you open your mouth and begin a slam-fest on someone you don't like in the media or go on the personal attack with someone you work alongside, remember that your little rant is reaching MP3 players around the world. Be sure — before you open your mouth to speak — that you aren't misquoting an article or merely assuming that your word is gospel. Cite your sources and make certain those sources are not only reliable and authentic, they are confirmed by other credible sources. If you're doing a news podcast or include a news segment in your show, consider citing the source where you got your news. It not only tells your listener that you're not making this stuff up, it's a nice way to drive traffic to the website you consider valuable. When expressing opinions, jaded, constructive, or otherwise, have real evidence to back up what you say — and put up or shut up!

WARNING

Also, keep in mind that your freedom to express yourself and speak your mind comes with responsibility and consequences. Your words carry weight, and if you go after others in your professional circles or in your community, there is no guarantee your listeners will back you 100 percent. You may picture your fans rallying around you and your show but find yourself working damage control instead. Before wrapping yourself in the First Amendment, remember it is a right, not a privilege.

Chapter **6**

Interview-Fu: Talk to Me, Grasshopper

odcasting is empowering. There's something about a microphone in your hand that gives courage. Suddenly, you're not afraid of anything. Oh yeah, you're "running with the big dogs" now, and like Paula Zhan, Stephen Sackur, or Anderson Cooper, you're asking the questions to find out what makes your guest tick.

What sets you apart from those big dogs, though, is skill. The late, great Anthony Bourdain may make it look easy on *Parts Unknown* in how people would open up to him over a local dish or an original culinary experience; but that was the magic of the perfect setting, the best dish, and the charm of Bourdain. He could click with people. (His interview with Danny Trejo is a favorite of Tee's.) Make no mistake: Interviews are not easy. There is a skill in hosting an interview, and hosting a great interview is an art. The good news is these luminaries blessed with the gift of gab all had to start somewhere. Podcasting is an excellent venue to develop and hone these skills, but you're going to want a solid foundation to build your skill set on.

Along with helping you schedule an interview, we help you get ready for it by looking at hardware and software tried and true for us, asking good questions to keep the conversation lively and engaging, and giving you examples of bad questions best avoided. Finally, we impart those always-valuable behind-the-scenes technical tips that make the interview go smoothly.

I'll Have My People Call Your People: Interview Requests

The courage to submit an interview request comes simply from your interest in the interview subject. Script or compose an email to ask your favorite author, actor, sports celebrity, streamer, podcaster, or whomever you want for an interview. You may need to submit the request multiple times, and sometimes you may have to work through numerous people simply to get a "no" as your final reply. That happens. It doesn't mean that individual is mean, a rude person, or otherwise. They just don't do interviews. For every "no," you will find ten others who will enthusiastically say "yes."

Here are some things you should keep in mind when working on the interview request:

>> **Market yourself and your show.** A good deal of marketing is involved with podcasting. Your interview request needs to sell your services to the prospective interviewee. If you're part of a podcasting network, be sure to mention that. Large listenership numbers are always helpful. Have you done interviews before? If so, do some name dropping. If not, a good place to start might be with other podcasters. They're looking to get their names out and grow their listenerships also.

>> **What can I do for you?** The person (or the person's agent) is going ask, "What's in it for me (or my client)?" You need to ask yourself questions like: Does he or she have a new book coming out? Perhaps he or she is about to launch a special product? In the case of *Technorama* and the Hoover Dam, it was a genuine interest in the science and the history behind the engineering achievement. That pitch granted Chuck and Kreg, pictured in Figure 6-1, an incredible behind-the-scenes look at Hoover dam. Find an angle and work with it.

>> **Be flexible.** Remember, you're asking for *their* time. There may be restrictions in your schedule and theirs. Sometimes you can get an interview within 24 hours and other times you have to schedule it weeks or months in advance. Sometimes, you get lucky and record an interview right on the spot. You may have to take time off work from your regular job, rearrange other plans in your week, or outright cancel things you have previously planned, just like the interview subject who is taking time out of their day to chat with you.

REMEMBER

Don't assume the person reading your interview request is going to know your podcast, or even what a podcast is. (Yes, we've been at this for over 15 years and there are people who still don't know what a podcast is.) You may have to explain your platform using alternative terms or a short explanation.

THE INTERVIEW REQUEST

One of the easiest ways to do interview requests is via email. It's time to put your marketing hat on top of your public relations hat and consider how you want to represent your show when sending queries. First, putting words like "Interview request" or "Podcast interview request" in the subject tells the reader what you want before they even open the message. Use a warm greeting: "Hello" followed by the person's name is always a good start. Remember, you may not be sending the message to the interviewee directly, but rather have to work through their agent or handler.

In the body of the message use the B.L.U.F rule — or Bottom Line Up Front. The first line of your message body should include something along the lines of "I am requesting an interview with Brent Spiner for *The Topic is Trek* podcast" — again, no doubt what you want. Don't dive in to the details about your show yet. We know, it's tempting right after you mention it by name, but remember, this query is about *them*. You can mention what your podcast is about with a one-liner about what you do ("*The Topic is Trek* is a show about *Star Trek* news, episodes, and thought-provoking questions in a fun and energetic tone.") or it may be all in the title ("We would love to host an interview with Shohreh Aghdashloo in an upcoming episode of *Happy Hour from the Tower: A Destiny Podcast*."), but right now, the interview subject is our focus.

Explain why you want to speak with that person. Do they have any upcoming/recent books, movies, or appearances they are interested in promoting? This would be a good time to mention how much time you are asking for. Asking someone for 2 hours is a much different consideration (and easier to say no to) than 15 minutes.

Now is when you can get on with describing your show, but don't drone on — a couple of sentences and perhaps some other notable names whom you have interviewed ought to be all you need. If you have an audio or video file that describes your show, include a link to that, as well.

Close with a respectful, yet positive salutation such as "I look forward to hearing back from you." Remember, most people are very busy and may not respond back right away so you might need to follow up periodically. In some cases, several times. It took Chuck two years of polite, yet gentle, reminders to get an interview with Dr. Robert Ballard (the man who discovered the *Titanic*), but it was well worth the effort.

FIGURE 6-1:
In 2008, the
Technorama crew
got an exclusive
behind-the-
scenes look at the
operations of
Hoover Dam after
submitting a
formal sincere
request for an
interview.

TIP

Don't forget to exchange contact information with your guest once your interview is confirmed. Will you be calling their phone or will you be using software like Skype, Facebook Messenger, Zoom, or Discord? This will have an impact on what you will need to record the call (covered later in this chapter).

Preparing for Interviews

There's an approach that all interviewers, be they Barbara Walters or Stephen Colbert, should take in talking to guests — use a simple, basic plan to ask the questions that garner the best responses.

Asking really great questions

Chances are good that if you're new to podcasting, you've never held an interview quite like this — an interpersonal, casual chat that could get a bit thought-provoking or downright controversial, depending on your podcast's subject matter. The interview may be arranged by you, or it may be prearranged for you. There's a science to it, and here are just a few tips to take to heart so you can hold a good, engaging interview:

>> **Know who you're talking to and what to talk about.** When guests appear on your show, it is a good idea to know at the very least the subject matter on

which you will be talking. Let's say, for example, you are having an author appear on your show. If the author has written over a dozen books, be they fiction or nonfiction, trying to find the time to read all of your guest's books would seem an impossibility. So do some homework. If the author guest has written a popular series, go online and research the series. Visit `Wikipedia.org` and see if the series has a summary there. If you can only find limited information, find websites relevant to the topic of the series. If the series is steampunk, dig up information about the Victorian era. If the series follows a snarky, sentient robot, look up Artificial Intelligence. This has two effects: (1) You sound like you have a clue what the writer is writing about and (2) it allows you to ask better questions. These same rules apply for nonfiction authors, and really for guests of any particular background.

TIP

It's also a good idea to visit guests' websites and social media platforms (provided they have 'em). You don't have to be an expert on their subject matter, but you should be familiar with it so you know in what direction to take the interview.

» **Have your questions follow a logical progression.** Say you're interviewing a filmmaker who is working on a horror movie. A good progression for your interview would be something like this:

- What made you want to shoot a horror movie?

- What makes a really good horror film?

- Who inspired you in this genre?

- In your opinion, what is the scariest film ever made?

You'll notice these questions are all based around filmmaking, beginning and ending with a director's choice. The progression of this interview starts specific on the current work and then broadens to a wider perspective. Most interviews should follow a progression like this, or they can start on a very broad viewpoint and slowly become more specific to the guest's expertise.

» **Ask open-ended questions.** To understand open-ended questions, it's simpler to explain *closed-ended* questions. Close-ended questions are the kind that give you one word answers — for example, "How long have you been studying plate tectonics?" Don't be surprised if your interviewee comes back with "I started my studies in plate tectonics seven years ago." And then . . . silence. Close-ended questions make the process harder than it needs to be. Instead, rephrase your question like "So what exactly got you interested in plate tectonics?"

» **Prepare twice the number of questions that you think you'll need.** Some interviews you hear grind to a halt for no other reason than the interviewer believed that the guest would talk his head off on the first question. You're certainly in for a bumpy ride when you ask a guest, "Tell the listener a little bit about your experience at WidgetCo," and the guest replies, "It was a lot of hard work, but rewarding." (Yeah, this is going to get painful.) This is why you should always have far more questions than you need. If you plan that every question merits a 2-minute answer on average, and you have 20 minutes with the subject, have 15 to 20 questions ready to ask. It's always good to have a few questions in your reserves.

TIP

Have a pad and a pen on hand, ready to go. In the middle of your interview, an answer may inspire a brand-new question you would want to ask your guest. Jot it down so that you won't forget it. Then ask this new question either as a follow-up or in place of another upcoming question.

» **Never worry about asking a stupid question.** When asking questions that may sound obvious or frequently asked, remember: Chances are good that your audience has never heard them answered before. Okay, maybe a writer has been asked time and again, "Where do your ideas come from?" or a steampunk, as seen in Figure 6-2, has heard, "So, how long did it take you to make this prop?" often. When you have a guest present for a podcast, there's no such thing as a stupid question; what's really dumb is not to ask a question that you think isn't worth the guest's time. He or she may be champing at the bit in hopes you will ask it.

FIGURE 6-2:
While interviewing steampunks at DragonCon, Chuck never worries about "How long did these props take you to make?" as the often-asked question is welcomed by makers, cosplayers, and other creatives.

REMEMBER

Leave room for spontaneous questions. Listen to your guest's answers and see if a new path has opened up. They may be tense while answering the same questions for the 50th time, but if you strike a chord and stumble on a piece they are passionate about, abandon the questions for a bit and follow the trail! It can lead to some extremely interesting conversations and stories.

Avoiding really bad questions

Before you start percolating and dream up a few questions based on the preceding tips, stop and think about the interviews you've listened to where things suddenly headed south. Usually the interviewer finds themselves with a guest they know nothing about and they are expected to interview them on the fly, or the host ambushes the guest with questions that dig into something that's out of the guest's scope or none of the interviewer's business. We've piled up the typical gaffes in a prime example of a good interview gone bad.

Every podcaster should know how to turn a pleasant conversation sour (uh, this *is* a satire and not a recipe, okay?), and the following blunders should do it faster than an Uwe Boll movie is in and out of theatres:

>> **Ask inappropriate questions.** Keep in mind your podcast is not *60 Minutes, HardTalk,* or even *Judge Judy.* If you want to fire off hard-hitting-tell-all-mudslinging questions, think about who you're talking to and whether the question is within the ability of the guest to answer honestly and openly. If not, an awkward moment may be the least of your worries. Inappropriate questions can also be those irrelevant, wacky, off-the-wall, and far-too-personal questions for your guests. "Who was the rudest person you have ever worked with on a set?" could put a stunt performer's career into jeopardy if answered earnestly. "What's the worst book you've ever read?" could drop a writer into hot water with their colleagues. Asking athletes "You are in fantastic shape. Do you sleep naked?" could easily derail an interview. Maybe these "wild card" questions work for shock jocks, but when you have an opportunity to interview people you respect in your field, do you really want to ask them something like, "Boxers, briefs, or none of the above?" Think about what you're going to ask before you actually do.

>> **Continue to pursue answers to inappropriate questions.** If a question has been deemed inappropriate by a guest, don't continue to ask it. Move on to the next question and continue forward into the interview. Podcasts are by no means an arena for browbeating guests into submission till they break down in tears and cough up the ugly, sordid details of their lives.

And news flash: They're not going to.

Are there exceptions to this exception? We would say, yes, depending on the content of your podcast. Say after reading — and enjoying — *Podcasting For Dummies,* you decide to become the Zach Galifianakis of podcasting, complete with foliage for a backdrop. (And yes, we had that in the early years of podcasting with Kent Nichols and Douglas Sarine's *Ask a Ninja,* archived at `https://www.youtube.com/askaninja`.) Of course, if you're after irreverent material for your show and push that envelope as far as you can, your guests may not want to play along — especially if they don't get the joke. If that's the case, expect your guests to get up and walk away. Even in the most idyllic situations, guests can (and do) reserve the right to do that.

>> **Turn the interview into the Me show.** Please remember that the spotlight belongs to your guest. Tee recalls a podcast where — no kidding — the three-person crew invited a guest on their writing podcast to talk about their books and their methodology of writing . . . only to launch into a 15-minute discussion between themselves on a completely unrelated topic, leaving the guest on the other side of their mic. Silent. *For 15 minutes.*

Yes, it is your podcast, but when a guest is introduced into the mix, you're surrendering control of the show to him or her. That isn't necessarily a bad thing. Let guests enjoy the spotlight; your audience will appreciate them for being there, which adds a new dimension to your feed. One way to avoid the "me factor" is to think of yourself as a liaison for the listener. Ask yourself, "As a listener, what questions would I ask or information would I be looking for from the guest?"

>> **Respect your guests. Period.** It has happened to Tee, both as an interview subject and as an interviewer. He's answered some questions that made him uncomfortable, and requested podcasters to please edit out the question and related awkward response. In most cases, the podcast respected his request. The others who did not bother to edit their podcasts? Well, he no longer fields queries from them. With that experience, Tee extends the same courtesy to his guests. Why? Interview subjects talk to their friends and you want them to speak positively of you and your podcast. Show them respect, and your guests will do the same.

Feelin' the synergy

One final note on preparing for interviews: We've heard some guests say, "I'm doing these interviewers a favor by going on their show." And we've been told by other show hosts, "We're doing you a great favor with this chance to showcase your work on our show."

Both of these opinions are very subjective.

The reality is that host and guest are working together to create a synergy. The interviewer has a chance to earn a wider audience and display mastery of journalistic techniques. The guest has a chance to get into the public eye, stay in the public eye, and talk about the next big thing he or she has coming in sight of said public eye. Working together, guest and host create a seamless promotional machine for one another. And if you are lucky, the interview subject might just take complete control of your show. That is what actor Lani Tupu did to Tee in a recording session captured in Figure 6-3. This reflected the trust as well as the fun the *Farscape* actor was having alongside Tee.

FIGURE 6-3: If a real synergy is created between interview subject(s) and host(s), you can ramp up the fun while recording a podcast, the prepared questions set aside while you riff on the mics.

If you decide to take on the art of the interview, keep these facts in mind; you and your guests will have your best chance to work together to create something special.

REMEMBER

If your format allows it, ask your guest for an ID that you can drop in from time to time. You've probably heard these before on radio stations "Hi, this is Rex Kramer, danger seeker. You may remember me from such films as *Airplane* and *Kentucky Fried Movie*, and you're listening to The Shameless Self-promoting Podcast." If the interview guests want to be more creative, let them. These are a great self-promotion tool, a whole lot of fun, and a way to remind your listener of previous accomplishments. Remember to ask politely, and be aware that not everyone will (or can) comply.

Recording Interviews

Unless you're conducting in-person interviews, your podcast just got a bit more technically complicated. You need to have the appropriate software to record your interview over the phone.

One of the most popular options of recording your interviews is with Skype (www. skype.com). What makes Skype appealing to podcasters is its expandability of the application, available for Windows, Mac, and Linux. Skype is the vehicle to make and record your calls. You can also use Skype with a little more hardware for similar results. We discuss all your options — whether software or hardware — in the following sections.

WARNING

There are legal restrictions concerning the recording of telephone/Skype calls, and these restrictions vary from country to country, state to state, and region to region. Compliance with these laws is the responsibility of the podcaster. Always ask for permission (or better yet, get it in writing) before recording phone calls.

Recording using Skype

Recording using software or hardware — once again you're faced with choices. Software solutions are typically less complex to set up and use and cost less; however, they can put more load on the computer's CPU. If the CPU is too busy, it could impact the quality of the recording.

Recording a Skype call is really very simple.

1. **Launch Skype on your device.**

2. **Call the person you would like to interview.**

    This is a good time to enter contact information and place the call. Let your guest know you will begin to record now.

3. **Click the menu in the lower right.**

    The menu choice is three horizontal dots, often referred to as *meatballs*. This menu is available on both mobile and desktop; however, it may not show up on the desktop application until you move your cursor over the Skype window.

4. **Choose Start Recording from the menu, as shown in Figure 6-4.**

    The person on the other end sees a message that the call is being recorded.

FIGURE 6-4:
Record Skype
calls with the
lower right menu.

5. **Use the same menu to stop the recording.**

   When you're ready to end your recording, use the same menu in the lower right and choose Stop Recording. If you disconnect the call, the recording will automatically stop so remember to restart another recording if you get disconnected and call back!

6. **Open the recording.**

   The recording is available in your Skype chat window. This is normally displayed after you end the call, or you can find it at any time during the call using the chat icon.

7. **Save the recording.**

   The recording begins to play. This is a good way to verify everything went as planned. Use the meatball menu and choose Save As to save your file. The file is saved as an *MP4*, a standard video format. If you just want the audio, ensure that your editing software can handle importing. We recommend doing a simple test with a friend before your big important interview to make sure that you don't end up with your masterpiece in a file you cannot use.

REMEMBER

Skype call recording is available for Skype-to-Skype calls. If you want to record Skype-to-phone (landline or mobile) calls, it requires additional software that goes beyond the scope of this book. A quick search of "Skype recorders" should turn up several options.

TIP

Skype uses *VoIP* (**Voiceover Internet Protocol**) to do what it needs to do in order to help you host and hold audio and video conferences. If you have a good Internet connection and want an all-in-one solution, consider looking into Zencastr (http://zencastr.com/) as a possible solution for recording interviews. It will require that your interview subject on the other side of the website has audio hardware hooked up on their end, but the end result is audio so cleanly recorded, you would think both host and guest are in the same room together.

Recording using OBS

If Skype-to-Skype calling isn't an option, then it's time to bring in another player: *OBS*. We talk a bit about *Open Broadcast Software* (OBS, found at https://obsproject.com) in Chapter 17, but we briefly touch on it here, using it as a virtual mixer for recording an interview. If you are thinking "Wait, I'm working with a mixer?" we've got good news: We cover a few affordable options and offer more details on mixers in Chapter 2. We're taking those principles and applying them with OBS.

REMEMBER

The idea behind a mixer is to take multiple inputs and mix them to create a singular output. Input sources can be a microphone, a computer, a MP3 player, an electronic keyboard, or just about anything with an audio output you feed into the mixer. With OBS, your input sources will come from any hardware or software connected to the computer. For this exercise, we employ another communications app, Discord (https://www.discordapp.com), to host our interview subject, and use OBS to record our interview subject.

WARNING

There are legal restrictions concerning the recording of calls, video or audio, and these restrictions vary from country to country, state to state, and region to region. Compliance with these laws falls on the content creators. Always ask for permission (or better yet, get it in writing) before recording.

Before going to Discord to bring in your guest(s), you're going to want to get set up on OBS first:

1. **Download OBS Studio from** https://obsproject.com, **install, and then launch the app.**

2. **Go to the Scenes window on the bottom-left of the UI and right-click on the default Scene displayed there.**

3. **Choose Rename and call this the Recording Booth scene.**

4. **Next to the Scenes window is a Sources window; click the + option to view the Source menu, as pictured in Figure 6-5.**

   OBS transforms your computer into a fully working recording studio, complete with multiple incoming sources that create a scene. We are creating a scene for our recording session in Discord.

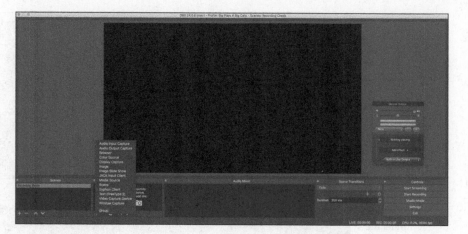

FIGURE 6-5:
Open Broadcast
Software (OBS)
transforms your
computer into a
fully working
broadcast studio,
complete with
multiple input
sources.

5. From the Source menu, choose the Audio Input Capture option.

6. In the Create New option, label it as Discord Capture and click the OK button.

7. Select the Default option and then click the OK button.

As OBS is recording your audio, you do not need to set up any additional input sources as you have already done this in Discord. OBS records the default audio, the particular input sources are handled by Discord.

8. From the Source menu, choose the Image option.

9. In the Create New option, label this as Background and click the OK button.

10. Click the Browse button and select an image from your hard drive.

11. Click the Open button to drop in the image and then click the OK button to return to the OBS interface.

Adding the background image is purely optional. We're doing this as we're exporting video from OBS once we are done.

12. Go to OBS ⇨ Preferences in the Application menu and choose the Output option.

13. From the window that appears, look at the Recording section and review the Recording Path featured.

14. Designate where your files are saved (see Figure 6-6).

You can either click the Browse button and leave the location on default or designate a new save location.

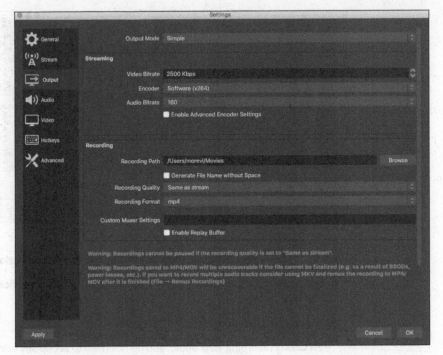

FIGURE 6-6:
Under the Settings window, you can designate where recordings made in OBS are saved as well as what video format in which your recordings are saved.

15. For the Recording Format section, click on the drop-down menu and choose mp4 as your recording format.

16. Launch Discord and set up your interview meeting.

17. With OBS and Discord running, go to OBS and click the Start Recording button.

18. Return to Discord and begin your interview; when you are done, single-click the Stop Recording button.

19. Go to the folder you designated as your Recording Path and find your recording.

Congratulations! You just recorded an interview using OBS.

Presently, the video file you have is your background image, and the interview is audio only. You can go into a video-editing program or a player as simple as QuickTime and export the audio from this video file. From here, you can take the audio and prep it for podcasting or whatever you want to do for your interviews. Again, in Chapter 17, we work with OBS more in depth; but if you want to know more about other capabilities of OBS beyond this exercise, take a look at *Twitch For Dummies* from our own Tee Morris.

And if you are wondering "This Discord thing is cool. Where can I find out more about it?" then take a look at Tee's other title, *Discord For Dummies*.

WARNING

If you're doing an interview with multiple participants (known as *conferencing*), it's often best to have the person with the highest power CPU host the meeting. Optimally, that is you. That person should initiate the call and invite the other attendees one at a time. The better the CPU, the better the conference will run, and the better your recording will sound. Also, the conference host may or may not be the same as the person recording the call. Remember, your goal during an interview is to minimize the chances for problems.

Recording using hardware

If recording using software has your head spinning, using hardware may simplify things. So instead of a virtual mixer like OBS, we're using a physical mixer. The input source can be a microphone, a computer, a MP3 player, an electronic keyboard, or just about anything with an audio output to feed in to the mixer. As seen in Figure 6-7, your input as the interview host will come from the microphone while your subject's audio will come from a computer running call-in software you are familiar with. The output needs a destination, such as a recording device like a Zoom H4n.

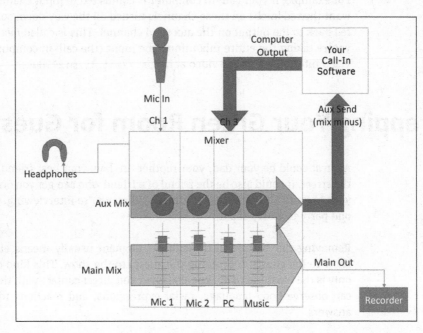

FIGURE 6-7:
If you prefer to work with hardware as opposed to software for recording, a mixer, a digital recorder, and a computer running call-in software can be implemented to record guests.

All right, we admit that while Figure 6-7 looks a bit challenging, it's not that complex. Whether you have a basic four-channel mixer or something the size of a cruise ship, most mixers have the same basic layout and features. Plugging in a microphone to a single audio channel is pretty straightforward. Adding the stereo output of a destination device (a second computer, a digital recorder, and so on) is also pretty simple, provided you have the right connecting cables. The real fun begins when you need to get your output back to the call-in computer and the recording device. This is where you need to take advantage of your mixer's *aux send* (sometimes called *effects send*) port. This is an output port that you will feed back to your call-in machine. So the question now becomes, how do you get your microphone input sent out the aux send?

Most mixers have the ability to create multiple mixes. The *main mix* is what you typically record, but often hidden mixers, called *buses,* let you create an alternative mix. How cool is that? You thought you were just buying one mixer, and you got yourself one (or more) for free! To make use of this additional output mix, locate the row of knobs on your mixer labeled *Aux,* often red in color. (If you have more than one row of Aux knobs on your mixer, each row corresponds to a separate aux send channel.) These are the volume controls for your aux send. If you turn up the volume on your microphone, whatever is connected to listen to the aux send output will hear it.

For example, if your call-in computer is connected to input channels 3 and 4, you want the red knobs on those channels turned all the way down so that input isn't fed back to the output on the aux send channel. This is called *mix minus 1* because you're taking the entire mix minus one input (the call-in computer). For a short video tutorial, watch the video at `https://bit.ly/mixminus`.

Prepping Your Green Room for Guests

A guest could be your dad, your mother-in-law, your best friend, or the man on the street. It could also be the friend of a friend who can get you on the phone with your favorite author, actor, or athlete. When you're interviewing, you have a second party to worry about.

Removing the "technical difficulties" element usually means either taking the show to the guests or bringing the guests to the show. This kind of interview not only is the most fun to do, but also gives you direct contact with the subject so you can observe body language, facial expressions, and reactions to questions and answers.

Welcoming in-studio guests

When you have guests visit your facilities — and because you're podcasting, this is probably your house — make them feel at home. Offer them something to drink. Offer to take them on a tour of your humble abode. Introduce them to your family. The point is to be polite. You don't have to cook dinner for them, but offering a hint of hospitality, be it a glass of water (or a beer, if you've ever worked closely with the *Binary Studio* crew), is a nice touch.

If you're having in-studio interviews, it's also a good idea to get your home and yourself ready to receive guests. Sure, Tee has recorded quite a few podcasts in his pajamas, but because he's working with his wife or recording short stories for his podcast, he's allowed. If fantasy and science fiction authors Elizabeth Bear and Scott Lynch ever come over to his house for an interview, don't think he'd be greeting them in his Avenger jammies and Stone Brewery slippers.

Okay, maybe he would greet them wearing the Stone slippers, but he would be bathed and dressed and have his teeth brushed and hair combed. The key word here is *guest*. Treat them as such. Be cool, be pleasant, be nice. And if you're a guest on someone else's podcast, the same rules apply. Don't prop your feet up on the furniture, don't demand hospitality, and don't be a jerk during the interview.

The in-studio visit is an audition for both guest and host. If the guest is abrasive, abusive, and just plain rude, chances are good that the guest will never be invited back, no matter how well the previous interview goes. If a host asks unapproved questions, continues to pry into personal matters that have nothing to do with the interview, or seems determined to take over the interview spotlight as if trying to impress the guest, said guest may never return, even if extended an invitation.

Meeting guests on their own turf

Be cool, be pleasant, be nice. These same rules apply when you take your podcast on the road. You may find yourself at a person's home, place of business, or some other neutral place. You're now practicing — for the lack of a better term — guerilla journalism, ambushing unsuspecting people with questions that may not strike you as hard and probing but could be to people who don't expect them. Make certain to show respect to your guests, wherever you are when the interview takes place.

A good approach for getting good interviews is to ask permission of your guests, be they passers by or experts at their place of business, to interview them. Shoving a microphone in someone's face and blurting out a question is hardly a great way to introduce yourself and your podcast to the world. If the guest you want to interview has a handler or liaison, it's good protocol to follow the suggestions and advice of the guest's staff.

If you start out with a warm, welcoming smile and explain what you're doing and why, most people open up and are happy to talk.

TIP

When interviewing people on the street or in the moment, there are some easy ways to identify yourself. Michael Butler of *The Rock and Roll Geek Show* (http://www.americanheartbreak.com/rnrgeekwp/) uses a *mic cube* around his microphone, also called *mic flags* (shown in Figure 6-8). The classic cube usually has a logo identifying a network, a show, or an organization affiliated with the interviewer. You can find mic cubes online (unprinted) starting around $25. There's also the simple greeting, "Hi, do I have your permission to record this for a podcast?"

FIGURE 6-8:
A mic flag is a great way to show your brand, either on camera or on the street, when interviewing people.

REMEMBER

Whatever your setup is, test your equipment before recording an interview. You're now out of the controlled environment of your home studio; you have to deal with surrounding ambient noise and how well your interview is recording in the midst of uncontrolled background variables. Set up your equipment; power up your laptop, mixing board, and mics; and record a few words. Then play back your tests and set your levels accordingly. When you have your setup running, you're ready to get your interviews.

Ensuring Trouble-Free Recordings

When it comes to recording conversations, here are a few points to keep in mind before asking the first question:

>> **Get permission to record conversations, even if the interview is prearranged.** Laws (both federal and state) prohibit the recording of conversations without permission, and further restrictions limit broadcasting these conversations. If you plan to record and publish a conversation, get the subject's consent (for both) beforehand, both verbally and in written communication (even email) to make sure your legal issues are covered.

>> **Test the calling equipment.** If you have arranged an interview with someone for your podcast — say, a favorite musician or politician — prepare for the interview ahead of time. Run a sound check and make sure the recording setup not only works but also sounds good.

 The bandwidth demand increases the more people you conference through your computer. Reception will be affected, so if you know more than one person will be involved in this interview, it's a good idea to test how many people you can effectively conference in one call.

>> **Check your batteries.** If you're using a portable recorder (such as the Zoom H4n Pro), make sure your batteries are charged and you have spares. (Check the spares, too.) If you're really paranoid or live in an environment with periodic electrical problems, you can also pick up an uninterruptable power supply in case your main power cuts out.

>> **Check your storage space.** Hard drives and solid state devices are getting bigger and cheaper, but that doesn't mean they're infinite. Audio files can be big — especially if you're recording to a raw format like WAV or AIFF! If you run out of space in the middle of recording a show or an interview, you lose time; lose pace; and in the case of interviews, lose face with your interviewee. If you're recording to a portable recording device, it's basically the same idea. Know how much storage you have, in megabytes or gigabytes, and how long you can record at your current bitrate. Don't worry; we talk more about bitrates in Chapter 9.

And although this may sound a bit pessimistic, be ready for things to go wrong. Guests might not show up for interviews. New high-tech toys, if not given a proper pre-interview shakedown, may not come through. Or sometimes, technology randomly fails. Prepare to have plenty of topics to discuss on your own. Have another recorder running as a simple backup. Plan for the worst, and that way, you can celebrate like a pro when that incredible interview goes off without a hitch.

IN THIS CHAPTER

» Setting your levels and parameters

» Focusing on volume and projection

» Capturing ambient sound

» Pacing and clock management

» Going off on tangents

» Recording

Chapter **7**

So What Are You Waiting For? Record, Already!

O kay, you've most likely gone through the hardware and software gadgets from Chapters 2, 3, and 4. In this chapter, we make the bold assumption that you've made your purchasing selections, hooked up all the hardware according to the supplied documentation (you did read the documentation, right?), chosen your software, and gotten everything ready to record. This is it! You have a microphone pointing in the direction of your mouth, eagerly waiting for you to begin podcasting.

Okay then. What's stopping you?

Perhaps you have no clue how you sound on your recording equipment, or perhaps you're still trying to understand why your smooth and sultry voice sounds like you just finished a dozen espressos. The problem could be in your audio application's sound settings. Too much pep in the voice, and you sound like Rob Zombie crooning a goth's delight. Too little amplification, and you'd be better off if you just went out to your front porch and shouted your podcast's content.

Before the podcast gets underway with your single-click or touch of the Record button, you must set *levels*. That's a very fancy-schmancy way to say *fiddling with knobs and sliders* on your mixing board or your audio-editing software. Setting levels ensure that the signal you're sending through the microphone is loud and clear. Once the technical side is running smoothly, give some thought to your voice — things like timing and enunciation — as how you say your message directly impacts the listeners' interest in what you're saying.

Did Your Sound Check Clear the Bank?

If you show up early enough for a rock concert, you see those roadies setting up microphones, playfully waving to the crowd as they speak quickly into a microphone "Check one, check two, check-check-check!" It's a staple for rock 'n' rollers to do such a mic check because the fans expect a good performance — it needs to be done.

With podcasting, your own mic check should be more involved. We recommend you perform an audio diagnostic just to assure yourself (and, if applicable, your guests) that the equipment is working and sound is, in fact, reaching your computer. The goal is not only to confirm that your mic is picking up sound, but also to check the volume of the voices — yours and that of whoever else is involved in this podcast.

Understanding dB levels

Setting levels is quite easy, provided you know where your *decibel (dB) input levels* are displayed on your software. The *decibel* unit is used to express the intensity of sound, beginning at -9 dB for the least perceptible sound to approximately 130 for an intensely loud sound level. Your readout measures how *hot* you are (in this instance, that's the power of your voice, not how good you look) on the microphone. Audio *signal strength* (measured in decibels) is the amount of power that goes into the signal, which affects how clearly it can be heard and how hard it hits the ears. "Loud and clear" is good; too much signal strength causes distortion, and that's a pain to listen to.

In vintage radio and audio equipment, a dB signal was represented in a *volume unit* (VU) meter — the little needle that bounced in response to your voice. Later on, the needle was replaced by lights that react to your voice or to sounds around you, going from green to yellow to orange to red. What the lights represent is pretty easy to translate:

>> **Green:** Well, I can hear you, but wow, are you quiet!

>> **Yellow-Orange:** You're coming in loud and clear.

» **Red:** You're in danger of hitting the distortion level (clipping).

» **Red with double bars:** Your audio is going to sound distorted.

Across hardware and software, volume meters appear different (Figure 7-1 shows the volume meter display for Audacity), but they all serve the same purpose: to make sure your content is heard clearly. Your aim, as you speak and watch the indicators speak back, is to keep your dB levels bouncing in the high level (low red) without lighting up those double bars. When you attain that average, your voice is rising and falling within a good balanced dynamic range. Try to keep things a little lower — in the red and orange range — with only the high points (when you raise your voice) going in to the red occasionally. You can always bring up the low spots with some simple post-production tips (described later in this chapter).

Input level meters

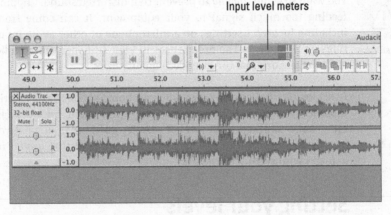

FIGURE 7-1:
The input level meters for Audacity (shown in the upper-right corner) respond to your voice and allow you to monitor how loud you are when recording.

Because microphones, audio-capture cards, and mixing boards all work at different sensitivity levels, it's best to test your voice before you record. If your levels are in the green, it means you're loud enough to be heard by your equipment but still so soft that people will have to crank up the volume on their computers or portable MP3 players, consequently blowing out their eardrums on the next podcast or playlist.

If your recording is too loud (or hot), it tends to cause problems for listeners. Listeners have their MP3 players set for a pleasant, comfortable volume . . . and suddenly your show begins with guns blazing and bagpipes blaring. As your levels reach deep into the red, listeners fumble for their players and try to turn down the sound so the program can be understandable through its own distortion. Sadly, the modulation is such a problem that your voice crackles and growls as it tramples the volume limits of your recording equipment, even at the lower volume. Even worse, some programs and microphones will even cut the signal from the mic for a second, leaving you with chunks of missing audio — and you may not discover this until after your precious recording is over!

CALLING IN THE CREW

It'd be great to set your levels so 0 dB is always where your lights remain, but recording is a real-world activity. Sometimes your meter may never reach yellow, and other times it might hit the double red bars. If your goal is to maintain perfect levels from the beginning to the end of a recording session, an audio crew can work with you in rehearsal and performance, adjusting levels when you go loud and when you go soft.

The downside of hiring a crew is an added expense to producing your podcast. Even if the crew is a collection of audio geek friends, it can still cost you when sending for a lot of pizza every time you record.

WARNING

You want to avoid *clipping* to prevent that distorted sound. Clipping is when you're feeding too much signal to your equipment. It can come from many sources including incorrect settings on a microphone or other audio input, incorrect settings on a mixer, or the preferences in your software. Once audio is clipped, you can't get the original (or intended) quality back.

In achieving a balance in your audio, you could spend the day setting and resetting those levels in quest of dB nirvana. That's time spent, but not well spent. It makes more sense to practice till you get a pretty good sense of what your best working level is and get comfortable speaking into a mic at that level. Time to go mano-a-mano with setting the levels.

Setting your levels

Your mission, should you choose to accept it, is to keep your levels in the neighborhood of 0 dB, dipping and spiking when necessary. For podcasting, consistency is key. You have your mixer turned on (if you have one), your mic is plugged in and turned on — check — and your software is running. Now follow these steps to check levels:

1. **Begin talking into the mic about your podcast topic or plans.**

    You can do a scripted test read or just talk off the top of your head, but be sure to speak in the manner and mood of your podcast.

    Instead of speaking to thin air in the vague direction of the mic (or the person next to you), speak in the exact direction of the mic (as shown in Figure 7-2). Even if the mic is omnidirectional (meaning it picks up sounds from all directions at once), it can pick up your voice better this way.

FIGURE 7-2:
Set your microphone at a comfortable distance, close enough to overpower ambient noise, but far enough to avoid microphone contact. Boom mic stands, like the one included with MXL's Overstream bundle, tend to be the easiest to adjust.

TIP

If you speak directly into the microphone, you might hear some sounds — particularly *ps* and *bs* — causing an effect known as *popping*. An investment of less than $20 in a windscreen or pop filter can fix that. This device is a foam or nylon filter that stops the wind made by your mouth from hitting the microphones diaphragm and causing distortion.

2. **While you're talking (or if you're monitoring by playing back your test takes), keep an eye on your dB levels on the computer screen.**

   If the levels are spiking into the double-red/red area or remaining in the green, check your input volume settings on your mixing board or audio-editing software. This might require a bit of multitasking on your part, but continue to adjust the input levels while talking, as you watch the input meters.

3. **Re-record your voice at the new settings.**

   Try to speak in the same manner and inflection as you did on the first recording.

4. **As you review the second take, watch the dB input levels and adjust accordingly.**

   When in doubt, err on the side of caution on the input levels. Record your volume a little low. You can always correct it in post-production. One way to increase the volume later, using Audacity, is to select your audio and use the Effect ⇨ Amplify feature to bring it up.

THERE'S A RIGHT WAY AND A WRONG WAY . . .

Time for a confession.

When Tee got his MXL990, it was his first studio condenser microphone. He was so excited about this microphone he failed to check the small slip of paper included that told him how to talk into it. Perhaps it was the excitement of venturing into new frontiers of creativity that made him hook up the equipment without reading the directions. Maybe it was a bold assumption that he knew what end to talk into with a mic as he had worked with (dynamic) microphones before. Whatever the reason, Tee was speaking into the microphone incorrectly (left image). His friends in the podosphere, on seeing Tee "in action," didn't really have the heart to tell him so — or didn't know any better themselves.

Fortunately, Evo Terra (https://www.evoterra.com/) is a heartless evil mastermind. He told Tee he was doing it wrong as any heartless evil mastermind would — on a podcast, naturally.

All microphones do not behave the same, and condenser microphones work best when speaking into them properly (right image), with the microphone pointed straight down. Tee, reflecting on Evo's "kind" advice, positioned the mic accordingly and continues to do so to present day. When hooking up microphones, refer to the enclosed documentation for additional and essential information on getting the best sound out of your microphone.

In short, understand whether your microphone is "end addressable," where you talk in to the end opposite the cable, or *side addressable,* where you speak in to the side; and you can avoid the embarrassment (caught forever) that published photographs may deliver upon you . . . regardless of how good your content is.

And when in doubt, read any documentation that's included with your new gear. It might save you some aggravation and a little embarrassment.

TIP

After you set your levels, make a note of the settings somewhere other than your computer (using, say, a smartphone or its retro ancestor — a legal pad and pen), in case your preferences are lost or fiddled out of whack by somebody's child or best friend playing spaceship in your studio. (Yes, even adults enjoy playing spaceship with really super-cool-looking equipment!) That way, even if something awful happens to your application's preferences or your mixing board, you always have your last known settings to reference.

Noises Off: Capturing Ambient Noise

Part of the charm that is podcasting is just how varied the content is, as well as how spontaneous the shows tend to be when the Record button is hit. Some podcasters believe that a "true" podcast (whatever that is!) must record everything in one take and deliver its content to listeners completely unedited. This supposed mark of authenticity includes any background noise (also called *ambient* noise) you happen to capture while recording — from comforting sounds like rustling trees and birdcalls to the more grating ones like pounding car stereos and jackhammers.

Hey, if that's the style of your show, that's great. However, if you're trying to take listeners to a place in their imagination, read on to find out how to reduce or eliminate ambient noise.

Identifying ambient noise

As we discuss in Chapter 8, how much you edit depends on what kind of content you're presenting. For example, if you're doing an off-the-cuff, off-the-wall podcast about your life and a typical day in it, you may just grab the VideoMic Me-L, plug it into your iPhone, and head out the door, recording every step along the way. This kind of podcast can be (note we said *can be*) easiest to record. You're podcasting a slice of Americana . . . or Britanniana, if you're in the United Kingdom . . . or *Kiwiana* if podcasting from the Land of Hobbits, Championship Rugby, and Pavlova. Especially if your goal is to capture the look and feel of your culture, ambient noise is not only welcomed, but desired. Up to a point.

Some podcasters cringe at the mere mention of ambient noise — ambient noise like . . . well, what was in Tee's very first podcast. When he podcasted *MOREVI: The Chronicles of Rafe & Askana* back in 2005, he worked to create a magical setting with voice, story, music, special effects . . . a world that was completely shattered by real-world interference like school buses, kids at recess, UPS trucks, the Virginia

Commuter Rail system, and air traffic from two nearby airports. Even if you love the source of the ambient noise, sometimes it just doesn't fit what the podcast is trying to do.

TIP

The best noise reduction happens in pre-production. If your podcast could do without input from the outside world, strategically scheduling recording sessions is the start of your production. Try recording at night or early morning. You'll find that traffic is lighter, the kids are in bed, construction crews aren't running those earlier-mentioned jackhammers, animals typically are less active — all adding up to less ambient noise, hence fewer takes on the mic.

Minimizing ambient noise

To reduce the intrusion of the outside world, record anytime during the day, and still maintain a budget, some creativity is in order.

Truth be told, there really isn't an easy solution to podcasting in a noisy world. One not-so-cost-effective answer is to rent a studio and record your podcast there. Unless you have a sponsor who bankrolls the costs, your "hobby" could easily max out credit cards and cast a hungry eye upon your nest egg. (Let's not even go there.)

A somewhat-less-expensive option is to soundproof your home-based recording room. That may sound simple, but it can involve a lot of home improvement before you have one room in which you can be sure the only sound is yours. But is it impossible or impractical? Not really. P.G. Holyfield of the podcast novel *Murder at Avedon Hill* (https://scribl.com/books/P4C82/murder-at-avedon-hill) built his own studio for podcasting.

Holyfield's do-it-yourself adventure began with a house hunt, so a studio in a finished basement was on the list of what the house needed. The studio eventually came about from a large storeroom, a few new walls, and the addition of an air vent. The newly created 7-x-8 foot room was then soundproofed with foam tiles. Holyfield found the effort worthwhile. "The studio is working out great. It was amazing to hear the difference as the foam went up. Now I have a completely silent room, except when the A/C turns on (which I can turn off most times, since the basement is pretty cool)." Figure 7-3 shows the process he went through. The advantage of being in the basement is far less outside noise entering through the walls. The advantage of the acoustic foam is to reduce sound waves bouncing off the walls and creating unwanted effects.

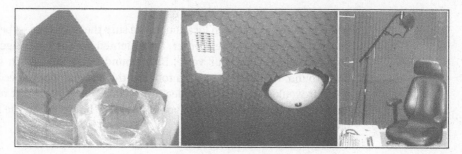

FIGURE 7-3:
P.G. Holyfield's
D.I.Y. studio
adventure: the
arrival of
acoustical foam
(left), measuring
twice and cutting
once (center), and
the studio (right).

Renovation, especially if you're a fan of D.I.Y. projects, makes this kind of home-built studio a possibility, but still may not be a practical solution for all homeowners — and this kind of aggressive home improvement is seriously frowned upon if you're renting an apartment.

You can keep this home renovation affordable and within the lease agreement terms:

>> **Stuff towels under the door.** It decreases the amount of sound from inside your house filtering into your recording area.

>> **Keep the microphone as far away from your computer as possible.** Its fan (if audible) simply becomes part of the natural ambiance for the podcasting room.

>> **Turn off any ceiling fans, floor heaters, additional air conditioners, or room ionizers.** With fewer appliances running, you have less chance of additional ambient sound being created.

>> **If you do encounter ambient noise that you don't want in your podcast, simply give it a few moments.** Wait until the noise subsides, pause, and then pick up your podcast a few lines *before* the interruption. That's for the sake of post-production: With a substantial gap in your podcast activity, you can easily narrow down where your edits are needed.

TIP

When noise interferes with your podcast, leaving gaps of silence so you know where to edit isn't exactly a foolproof method. Always set aside enough time to listen to your podcast — and really listen, not just play it back while you clean the office or call a friend. Make sure levels are even, no segments are repeated, and the final product is ready for uploading and posting.

Many podcasts rely on ambient noise to set a mood, but sometimes reality just doesn't cut it. If you want to put more craft into the setting for your podcast, the ideas in this section should help you keep the background down to a dull roar.

WARNING

When holding podcasts on location, make sure the ambiance — be it a particularly busy crosswalk at a street corner, a frequented bar, or backstage at a concert — does not overwhelm your voice. Background noise belongs in (well, yeah) the background. Its intent is to set a tone for the podcast, not to become the podcast itself. It would also be a good idea, if possible, to do a few test recordings in the space to see how much volume you'll need and how close to the mic you have to be in order to be heard.

Now Take Your Time and Hurry Up: Pacing and Clock Management

Podcasts, whether short and sweet or epic and ambitious, all share something in common: the need for *pacing*. As a rule, you don't want to blurt out the aim or intent of your podcast in the opening 5 minutes and then pad the remaining 10 or 15 with fluff. Nor do you want to drone on and on (and on . . .) till suddenly you have to rush frantically into why you're podcasting on this particular day about this particular topic. Give yourself ample time to set the mood comfortably and competently and make it to the intent without dawdling. Enjoy your podcast but respect your listeners' time. Make certain you don't overstay your welcome. The trick in pacing is to understand how much time you have to get your message across.

The big question is how to grasp how much time you really have. Where to start?

It's a good idea to get a grasp of how much time you really have in one whole minute. To do this, find a clock, watch, or a stopwatch and for one minute (and *only* one minute) sit quietly and do nothing, say nothing, and remain perfectly still. That one minute will feel like a short eternity. Now imagine that one-minute times 15. (No, no, no — do not repeat this exercise for 15 minutes. There's a fine line between an exercise and a complete waste of time.)

Fifteen minutes (for a start) is a good amount of time on your hands, so you should make certain that you take your time to get to the message of your podcast. The journey you take your listeners on doesn't necessarily have to occupy all of that time — be it 15, 20, or 30 minutes — but you have time to play. That's the most important thing to remember as you set your best podcasting pace.

Take the potato out of your mouth and enunciate

It isn't out of the ordinary to fire up the mic and launch into your podcast with enough energy and vitality to power a small shire in England. Nothing wrong with that — until you go back and listen to yourself. Your words, phrases, and thoughts are running together and forming one turbulent, muddy stream-of-thought. Yes, that is your voice, but even you are having a tough time understanding what the podcast is about and what the podcaster is saying . . . and you're the podcaster!

One way of getting a grip on pacing yourself through a podcast is *enunciation* — pronouncing your words distinctly, explaining your topic clearly — and there is no better way to do that than to slow your speech slightly and listen carefully to certain consonants (for example, the *t*, *s*, and *d* sounds). Proper enunciation can help you set a comfortable pace and keep you from rushing through your presentation. In fact, it isn't a bad idea to over-enunciate. In the excitement of recording, over-enunciation actually forces you to slow down, clean up your pronunciation, and make your voice easily understandable. That same excitement is likely to increase your pace of speaking. Again, if the rapid pace is part of your presentation, stick with it, but if you're doing anything like a narration for an audio book, slow down. If you think you're speaking just a little too slowly, you're probably going just the right speed.

TIP

Speaking a few tongue-twisters before going on mic helps your enunciation and also warms up your voice and prepares it for the podcast you're about to record. Here are some easy ones that emphasize the problem consonants lost when speaking too quickly:

>> Toot the tin trumpet, Tommy, in time.

>> Pop prickly pickles past the peck of parsley.

>> See sand slip silently through the sunlit seals.

>> The sixth sheik's sheep is sick.

>> Will Wendy remember wrecking the white rocker?

If you perform a search online for tongue twisters, you can find a wide variety of classic and original warm-ups for the voice. You can also take the twisters you find and compose your own.

And now let's take a break for station identification

Ever notice that certain podcasts suddenly break in with a show ID or take some other marked break and then (if they're lucky enough to have sponsors) play their sponsors' spots? They're trying to emulate — and maybe make themselves easily syndicated by — conventional radio. You know, the broadcasting industry. *Clock management* is the careful, standardized interruption of a typical radio broadcast — usually every 15 minutes in a one-hour slot — for such necessities as station identification, show identification, public-service announcements (PSAs), and (of course) commercials.

Another nice bonus of following clock management is you have several points in your show to give a quick *show ID* (something like "You're listening to Technorama with Chuck and Kreg . . .") that lets people know who you are. Chuck and Kreg like to collect audio clips from celebrities for their show IDs because, as seen in Figure 7-4, astronauts are cool! This show ID is a great way to catch the attention of people who may be walking into the middle of your podcast, and just as they ask the podcast listener "What are you listening to?" the show ID drops. Although show IDs are usually expected at the top of the hour and on the half-hour, you can drop in a show ID anytime you feel like it.

FIGURE 7-4:
Always have your recorder ready! Video game developer and astronaut Richard Garriott graciously gave Chuck an audio ID for his podcast at Dragon Con 2012.

A good general rule for clock management for each hour is

» :00 — programming break, station identification, show ID, PSAs, advertisements

» :15 — programming break, PSAs, advertisements

» :30 — programming break, station identification, show ID, PSAs, advertisements

» :45 — programming break, PSAs, advertisements

These breaks vary in length but can last anywhere from 30 seconds to 5 minutes. No surprise that this is where the old-school podcasters get a bit restless. They find clock management and show ID redundant, if not frivolous. Why bother with a show ID or clock management? People have downloaded your feed specifically. They know what they're listening to, so what purpose does clock management serve?

It all comes back to what you see for your podcast's future. If you intend to remain in the podosphere, it really comes down to your format, your way. No constraints from the FCC. It's just you, the microphone, your listeners, and throwing caution to the wind. However, if you set your sights on taking your passion project to the airwaves — be it AM, FM, satellite radio, or online streaming — you have to format your show to fit the standards.

Think about the possibilities of your podcast and where you see yourself with it several months (or years) down the road. Maybe a bit of management and organization will aid your podcast in jumping to the airwaves.

PODCASTING GOES PRO

Broadcasting networks like CNN, FOX News, BBC, and others often take their shows to the podosphere as a way to market their shows and reach a worldwide audience. Along with the talk-radio personalities, other major players in broadcasting like NPR, ESPN, Disney, and ABC use podcasting as a way to reach audiences that may have missed their original broadcasts. One of Chuck's favorites is a show from "the other ABC," the Australian Broadcasting Corporation, called *Dr. Karl on TripleJ,* while Tee enjoys NPR's *Note to Self* and Mike Rowe's *The Way I Heard It,* both of which were recommended to him by Stitcher. Thanks to podcasting, all of us can learn more about bee stings, double rainbows, and tidal waves on our own schedules.

Concerning Tangents and Their Val — Oh, Look, a Butterfly!

It's hardly surprising that broadcasters who have no experience in broadcasting tend to stray from their topic once the microphone goes hot, and then they lose themselves in the thickets and thorns of tangents. Defensive podcasters claim that's part of the charm of podcasting, but that charm fades as the topic gets closer to serious. If (for example) you launch a podcast dedicated to *The Witcher* series, but wander into a discussion over the growing concern of television violence, don't be surprised if your audience wanders away.

REMEMBER

It's worth repeating: Stay focused on the intent of your podcast. Remaining true to your podcast's subject matter isn't just about staying within your running time; it also makes clear to your listeners that, yes, you have a message and you will deliver it. You haven't promised something substantial only to let your mind and commentary wander aimlessly. When people listen, you want them to feel assured that what they hear is exactly what you've offered.

Read on to find out how you can make tangents work and how to smoothly get back on topic after you've taken your side trip.

"Say, that reminds me of something . . ."

Tangents can be creative opportunities. You don't necessarily treat tangents as strictly *verboten* in podcasting. If your tangent is directly related to, or benefits your topic, it can be effective and engaging.

Say your commentary begins with "Anyone notice how smartphones are no longer just phones but tiny personal digital assistants that can be easily monitored and hacked into?" You cover that for a bit, and then break off on a tangent about phone etiquette and the lack of manners it brings out of people. This is a tangent that will keep your listeners engaged, and that can count as much as staying on topic.

What if your podcast isn't so structured, though? There's nothing wrong with hitting Record and forging ahead into the great unknown of the next 15 or 20 minutes, so long as you have an idea of where you want to go. Tangents are terrific in moderation, but keep them at least (well, yeah) tangentially related to the subject matter of your podcast. If, say, your podcast is about movies but your review of *Doctor Strange* suddenly goes into the decline of the comic book industry (regardless of the onslaught of comic-books-to-film productions), then you still have a sense of focus. However, if your review of *Doctor Strange* wanders off into how your cat is throwing up hairballs every time you podcast or that your car's sunroof chose to stop working, blah, blah, blah, what the heck is going on here? Your audience may grow frustrated enough to stop listening.

"But getting back to what I was saying earlier . . ."

When you're podcasting for 15 or 20 minutes (or longer . . .), it's okay to take the scenic route with your discussion, but make sure that you return to the point you wanted to make in this particular (weekly, biweekly, or monthly) installment. The listener's delight in podcasts is often the revelation that individuals who (in many cases) have never set foot in a recording studio can produce entertaining, and even informative, shows. Sometimes meandering back to the point is part of that delight.

For instance, a podcast might begin with something read in the headlines — say, a new marketing strategy launched by Apple for the iPhone to reach a wider client base. This can lead to a variety of topics in the discussion, such as the following:

>> Smartphones that begin extending into other functionality like playing games on the PlayStation Network or providing Wi-Fi hotspots

>> Costs for iPhone services from AT&T, Sprint, and other carriers

>> Ways that carriers could bring down the costs of services

>> Continued shortcomings and disappointments from the smartphone industry

But in the last five to ten minutes, a simple segue like "So, to recap our thoughts on this bold marketing strategy from Apple. . ." can successfully steer you, or you and your co-hosts, back to the aim and intent of your podcast. If you return deliberately from a tangent, your listening audience arrives back at the point and knows the destination as well as the scenery they went through to get there.

Get your listeners there and back again. Not only will they appreciate it, but they'll tell others about your podcast, and your subscribers will grow in numbers.

We hit on the topic of using a script versus an outline in Chapter 5. When it comes to staying on track, clearly, a script has a distinct advantage over an outline. Regarding interviews (see Chapter 6), having a list of questions is important, but don't be afraid to ask impromptu questions — you can always go back to your prepared list.

THE NAME SAYS IT ALL

Unless you are already an established brand like Joe Rogan or Ricky Gervais, your show name is your calling card. For example, *Astronomy Cast* (http://astronomycast.com) — the name tells you right away what the show covers: galaxies, stars, gravity, black holes, and extra-solar planets. Once you have a topic for your show, it's extremely beneficial that the name conveys what the show is about. That way, when someone asks you *"So what's your podcast about?"* and you respond with *"Bite Me!"* they will understand it wasn't an insult as you explain, "It's a podcast about canines with temperament issues."

But give your show's description a few seconds. Just in case you want to mess with them.

Time to Push the Big Red Button!

Whew — there certainly is a lot to think about before you get started. That's what makes a great podcast — someone did some planning. Now it's time to put that planning and practicing into action.

Getting started with GarageBand

When you first start GarageBand, you're asked to start an Empty Project. A simple Audio project where a microphone is the main input source (pictured in Figure 7-5) is an available option. Selecting this option populates your GarageBand with a single audio track with a variety of vocal effects, some are practical if you are doing dramatic presentations, while others are comical effects for the fun of it. You can choose to utilize any of these effects and edit them using the GarageBand interface to cater to your needs.

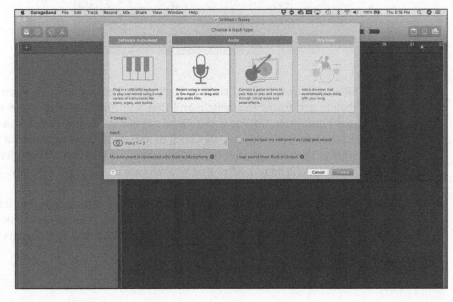

FIGURE 7-5: When starting off a podcast recording, GarageBand asks what kind of project you want to create. Single-click the microphone to set up your podcast episode.

We recommend changing a few settings that work better for music composition than they do for podcast recording:

>> **Turn off the Count-In and Metronome feature.** The Count-In sets a tempo and is most often used when recording music, but as you will probably not be concerned with a set tempo, you probably don't need this feature. After turning off the Count-In feature, turn off the metronome. If you don't turn it off, you'll hear a click-click-click while you're speaking. Annoying, to say the least.

>> **Hide the Library.** By default, a window will be made available on the left-hand side called the Library. The Library is a collection of effects for different kinds of recording. To hide the Library, click the Library icon, located in the top-left corner of the application window.

>> **Show Time in the LCD.** In the top-center of the GarageBand window is an LCD showing Beats & Measures. For your podcast, time is more appropriate so switch it by clicking on the music note/metronome icon. You will get a drop menu that offers the Time option (see Figure 7-6). Select that and your LCD now measures in time.

FIGURE 7-6:
Select Time to
measure in time.

Now you have a clean slate to start with. There's just one more thing to check before you throw the switch, and that's to confirm GarageBand knows where to get input from. Choose GarageBand ➪ Preferences and click Audio/Midi at the top of the window that pops up. Check to make sure your Audio Input is selected correctly. The setting varies depending on your actual input (USB microphone, mixer, preamp, and so on). After that's taken care of, close the window.

Return to your project window, click the big red Record button located at the top of the screen, and begin speaking. When you're ready to stop, click the Stop button or the spacebar. Note, clicking the Record button again stops recording, but GarageBand continues playing.

Congratulations, you've made your first recording in GarageBand!

Getting started with Audacity

Audacity was meant to be simple. To record a podcast in Audacity, follow these simple steps:

1. **Set your input.**

   At the top of the screen you can find a drop-down list with several choices —
   Line In, Microphone, and so on. Your settings depend on your system setup.
   Audacity is pretty good at detecting what inputs it can use on your system. Go
   ahead and pick the proper one. Just make sure that your microphone is
   plugged in before launching Audacity: Otherwise, it won't know your mic exists.

2. **Click the Record button.**

   Audacity creates a new track and begins recording. As you speak your words of wisdom to the world, you should see the meters moving and the waveform being created.

TIP

Audacity comes with a terrific function called Noise Reduction which digitally attempts to identify the offending noise and remove it from the recording. To do this effectively and efficiently, though, Audacity needs to have a clean sample of the room's natural audio, called an *audio floor* or *noise floor* in professional settings. Record 5 to 10 seconds of your recording environment, either at the beginning or end of your recording. Hit Record and leave the mics open but say nothing for a chunk of time, and that is the sample of your *room noise*. Use this clip when you are working with the audio in post, and Audacity's Noise Reduction asks for a sample of the background noise you want to eliminate.

3. **Click the Stop button.**

   When you're done, just click Stop.

Congratulations, you've made your first recording using Audacity. Now you know why so many podcasters love it!

And just like that, you're recording a podcast.

Chapter **8**

Cleanup, Podcast Aisle 8!

Although a high-tech activity like podcasting doesn't exactly qualify as quaint, the charm of podcasting is — whether produced by a major studio or by friends with a microphone and a laptop — its inherent, homespun quality. Steve Jobs once described it as the *Wayne's World* of radio. (No kidding, he said that in 2005 at http://bit.ly/jobs-podcasting on ABC.) Podcasts, much like *Saturday Night Live*'s classic sketch depicted, are often done on a shoestring budget and recorded in one take with no editing. All the trip-ups and tangents are captured for posterity and sent to MP3 players everywhere. This is part of the grassroots appeal that podcasting is not only known for but prides itself on nurturing as new and innovative podcasters enter the podosphere.

Although podcasting purists may harbor animosity toward the editing process, taking time to review and polish your podcast can truly make the difference between a listenable podcast and an incoherent mess of audio. For example, you might want to eliminate the sound of a train going by or silence a cough that would otherwise distract your listeners from a brilliant riposte. This chapter shows you how to use editing to shape your podcast while retaining its natural atmosphere, adding depth to that atmosphere with music, and then give the final touch to your podcast's format with an introduction and exit.

A Few Reasons to Consider Editing

Take a serious look at the mood you want to convey with your podcast. From there, you can judge how intense your editing workload will be. The following explains some instances where editing is needed.

Trust us. Your podcast will only benefit from it:

>> **Professional production quality:** You can't always get everything right on the first take — and sometimes not even on the second or third. Editing makes it sound like you got even hard-to-pronounce names and tricky tongue twisters right on the first try. What's more, you can drop in pre-produced clips with just the right amount of space before and after to give it that polished feel.

>> **Removing boring material:** You've probably watched a live show or listened to live audio that doesn't go quite where it is intended. The content gets dull and uninteresting, or everyone runs out of things to say. The only thing worse than aimless rambling are those deafening seconds of *dead air*. As a listener, you can fast forward, change the channel, or just turn it off. As a podcaster, you can avoid this situation with editing. But even though it sounds easy to take out the boring bits, be careful that you maintain continuity. If edited improperly, the listeners may find themselves confused because the line of questions changed from space travel to herbs and spices in the matter of a few seconds.

>> **Ambient noise:** As explained in Chapter 7, *ambient noise* is the natural and spontaneously occurring noise you may pick up when recording your podcast. If you're recording a story or setting the stage for a historic drama and suddenly a passing siren or the rumble of a garbage truck makes it into your recording, the odd noise can distract the audience. In fiction podcasts, such as *The Raven & The Writing Desk* (http://chrislester.org) hosted by author Chris Lester, moods and atmospheres must be maintained. Therefore, control and minimization of ambient noise is a must.

>> **Running times:** You just wrapped your latest podcast with a great interview, and you're confident that you have plenty of material for your 30-minute podcast. And then you check again and realize that you have recorded over 90 minutes' worth of interview. And you love all of it! Now here's where editing works in your favor. Your listeners expect 30 minutes, give or take a segment or two, from your podcast. You could run the whole thing, unedited, but that might test the patience of your audience members (not to mention your own bandwidth and file storage). Or you could split it up into two 45-minute interviews — or even three 30-minute interviews — breaking up the airplay of the interviews with two smaller podcasts in-between the segments. In this approach to editing, everyone wins.

Editing can easily increase your productivity with podcasts. True, some podcasters define editing as cutting and deleting material, but there's more to it. Editing can actually help you rescue discussions and content that would otherwise be hard to shoehorn into one podcast.

» **Scripted material:** Some podcasters argue that the true podcast is done in one take, but as podcasting matures as a medium, audiences grow more and more demanding. Expectations for a podcast change when your podcast includes scripted material. Editing is a necessity in these situations. With dramatic readings and productions, moments of "ah" and "um" should be edited out to maintain the clarity of the story, as well as maintain the mood or atmosphere established in your reading.

With the popularity of storytelling in the podcast universe, professionalism and performance are the keys to a good product — and usually that means (yep!) editing. Lots of it. If you feel that editing would mar the spontaneity of your podcast — but still want to present literature or other scripted material — ask yourself how good you think *Avengers: Endgame* would had been if Josh Brolin and Robert Downey, Jr. did that moment in one take. Imagine their dialogue sounding like this:

> *Thanos:* I am . . . uuuhhhhhh . . . inevitable. Yeah.

> *Tony Stark:* Ummm . . . and I *sneezes suddenly — sniffle sniffle* . . . am Iron *ahem* Man.

Not what we would call riveting. With scripted material, editing is a must.

And if you want to throw this book across the room, screaming "These jerks just spoiled *Avengers: Endgame*!" Come on, it came out in 2019.

Come. On.

The Art of Editing

Editing out breaks, stammers, and trip-ups may sound easy, but there's a science to it. If you cut off too much from a clip, one word comes right on top of another, and you sound unnatural. If you don't cut off enough, pauses last too long between thoughts.

REMEMBER

When you're editing audio, the key is to review, review, review.

A lot of audio applications are out there, each with its own way to edit stumbles, bumbles, and moments of silence to honor a lost thought. The same principles apply to all those applications:

1. **Find the unwanted content.**

2. **Give your clip a little bit — perhaps a half a second — of silence as leader or play area between edits.**

3. **Review the edit, making sure it sounds smooth and natural.**

   An effective edit doesn't sound like an edit. Any change of your audio should be unnoticeable.

But instead of talking about it, how about you do it? In the following sections, we go into basic editing using GarageBand (https://www.apple.com/mac/garageband/) and Audacity (https://www.audacityteam.org) as examples. These are two very common audio-editing software packages in podcasting, and both serve as our benchmarks for how to create podcasts. If you're using some other audio-editing software package, the steps are similar enough for you to apply to your own project.

Editing with GarageBand

You can use GarageBand to edit awkward gaps of silence or eliminate coughs and stammers from your podcast. We assume at this point you just finished recording or opened a GarageBand file with a recorded track already done (see Chapter 7 for instructions on recording with GarageBand). You can also import audio files (perhaps recorded on a portable device) by dragging them from the Finder into GarageBand.

To prepare for editing the silences and splutters, split your audio into smaller segments to isolate the audio that you want to remove. Then follow these steps:

1. **Determine where in the track you want to make the edit by clicking and dragging the playhead to the beginning point of your edit.**

   Use the Time Display readout to see exactly how long the gap of silence runs. The playhead tool is the triangle connected to a vertical line. (See Figure 8-1 for details.) When creating segments, you want to place this line at the beginning and ending points of your edit.

   TIP

   Silence is pretty easy to see when you're looking at a waveform of the audio. Coughs or other noise can be a little trickier to isolate. We suggest watching the waveform while you listen to the audio at the same time. You may want to use the slider in the bottom left to increase your time scale resolution and make it easier to work with smaller bits of information.

2. **Choose Edit ⇨ Split Regions at Playhead to make the first cut or use Command + T to perform the edit.**

Figure 8-1 shows the first cut.

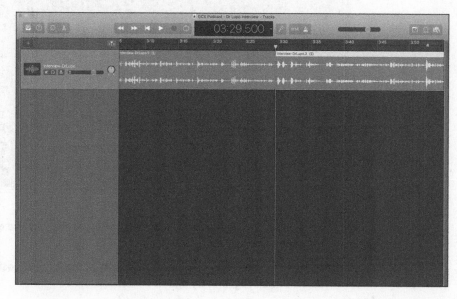

FIGURE 8-1:
Click and drag the playhead to the beginning point of your edit. The Time Display gives you an idea of where you are in the project's duration.

3. **Move your playhead to the location where you want the edit to end; then choose Edit ⇨ Split Regions (or Command + T) at Playhead to make your next cut.**

TIP

As you edit, give yourself a second or two of silent play area at the beginning and end of your edits. It makes the editing of two segments sound like one continuous segment, and they'll be a little easier to mix together.

4. **Single-click the segment between your two cuts and then press the Delete key.**

TIP

If you're working with GarageBand for the first time, note that the first track of audio is selected by default. To deselect the segment you're editing, single-click anywhere in the gray area underneath the track(s) you're working on.

5. **To join the two remaining audio segments, click and drag the right segment over the left segment, as shown in Figure 8-2.**

WARNING

If the selected segment overlaps any part of the unselected segment, it takes priority over the unselected segment, effectively erasing any content there.

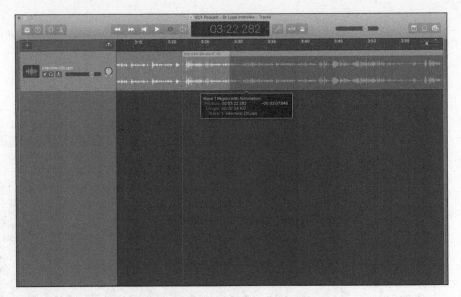

FIGURE 8-2:
To shorten the
silence, move one
edit over another.

6. **Click and drag the playhead to any point before the edit and click the Play button to review.**

7. **If the edit doesn't sound natural, undo the changes (by choosing Edit ⇨ Undo Drag) and try again.**

    Because you're allowed multiple undos in GarageBand, you can step back in your project to begin at the first Split command.

As much as we would love to cover all the neat doodads in GarageBand — and there are a lot of them — we need to stick to the essentials you need to get started podcasting. For more in-depth information on mastering GarageBand, there are plenty of other resources available, including Apple's own Support site at https://support.apple.com/garageband.

Editing with Audacity

At this point, we assume you've just finished recording something awesome with Audacity (see Chapter 7 for instructions), or perhaps you've opened a saved project. If you have an audio file from another source, such as a portable recorder, you can import it using File ⇨ Import. Now you're ready to follow these steps to make a basic edit:

1. **Find the segment that you want to edit.**

    To view the entire timeline of your project, click the Fit Project in Window tool, shown here in the margin.

TIP

You can easily navigate between the segments of your timelines selected and the project timeline as a whole with the Fit Project in Window and Fit Selection in Window tools, located in the top-right section of the project window.

2. **Click the Selection tool — in the upper left (Mac) and upper center (Windows) — and then click and drag across the unwanted segment.**

The unwanted segment is highlighted, as shown in Figure 8-3.

TIP

Right about here would be a good opportunity to use the Fit Selection in Window tool to double-check that the selection border doesn't go into the recorded content you want to keep.

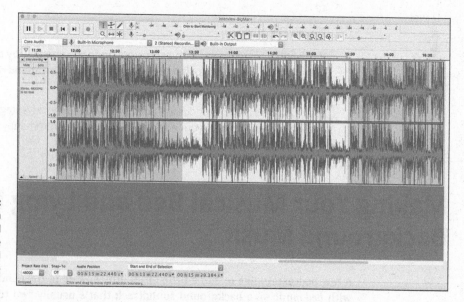

FIGURE 8-3:
With the Selection tool, click and drag across the unwanted content in the timeline.

3. **Single-click the segment between your two cuts and then press the Delete key (Mac) or the Backspace key (Windows).**

You can also choose Edit ⇨ Cut or press Command+X (Mac)/Ctrl+X (Windows) to remove the unwanted segment.

WARNING

Do not use the Trim command. (Here's the command not to use: Edit ⇨ Remove Special ⇨ Trim Audio or Command+T for Mac/Ctrl+T for Windows.) You may think you are trimming away unwanted content, but this command works differently from the Command+T command in GarageBand: Just like when you crop an image, it trims off unselected material — leaving you with only the content you wanted to edit out. Ack!

4. **Review the clip.**

5. **If the edit doesn't sound natural to you, undo the changes by choosing Edit ⇨ Undo Change of Position Region or pressing Command+Z (Mac) or Ctrl+Z (Windows), and try again.**

   You're allowed multiple undos in Audacity, giving you the advantage to go back to the beginning point of your editing just in case you aren't happy with the sound of the edit.

Making Your Musical Bed and Lying in It: Background Music

A few podcasters like to add a little bit of atmosphere in their individual podcast with *bed music* — a background soundtrack that's usually two to three minutes long and is looped so it can play again and again throughout the podcast, if desired. Sometimes the bed music lasts only a minute or two into the podcast when the hosts return from the break, whereas other shows keep it going from beginning to end, fading it out if they're bringing in any other sources of audio (voice mail, other podcast promos, ads, and so on).

REMEMBER

Regardless of how long the loop is, you must obtain permission from the artists to use their music and give them audible credit — either at the beginning or end of your podcast. If you're using shareware music, such as from Freesound (https://freesound.org), giving audible credit is one of the conditions you must agree to. Other options include using music loops found in GarageBand, Audition, or Logic Pro.

Finding the right balance

As you add music as a background bed, incorporate sound effects, and bring in pre-recorded audio from other sources (such as H4n Pro recordings or a Skype call), the sound of your podcast gets more complex. Balance becomes not only harder, but even more essential. (Note, for example, the various volume levels of the multiple tracks in Figure 8-4.)

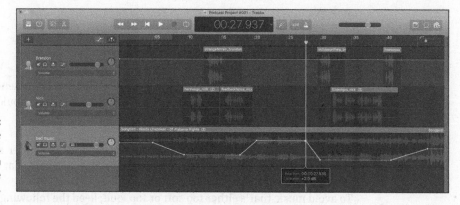

FIGURE 8-4:
Adding multiple tracks is easy. Balancing them can be challenging.

Bed music should add atmosphere to your podcast. That means finding a proper balance between the talking of the hosts and the soundtrack. The following sections explain why you want to avoid music that's too soft or too loud.

What is that noise?

If your music is too soft, your looped music could be mistaken as unwanted ambient noise of someone listening to music in the next room. Or it could be regarded as some kind of technical difficulty such as a stray *signal bleeding* (another wireless audio signal accidentally being picked up by the same frequency as your own wireless audio device) into the podcaster's wireless microphone. Music too indistinct to hear can be a distraction, particularly in quiet moments or pauses in the conversation. Your audience might end up hammering out email after email (asking what's making that annoying noise) or trying to figure out what that faint music in the background is.

Could you speak up? I can't hear you for the music . . .

When bed music is too loud, your voice is lost in the melody. Music, especially classical music and selections that rise and fall in intensity (like any good Queen album), can be really tricky to mix into a conversation.

You do want to allow your audience to hear the music in the background, but the music, if you're using it as background music, probably isn't the point of the podcast. It shouldn't be so loud that you have to pump up your own voice track to be heard.

Never sacrifice audio clarity so bed music can be heard and identified clearly. If you want to showcase music, then showcase music properly. Otherwise, your bed music should remain in the background as a setting, not in the forefront of your podcast.

Applying bed music the right way

When you're setting audio levels, you want to find the best blend of music and voice, assuring one doesn't overpower the other. Both tracks should work together and not struggle for dominance.

REMEMBER

Always listen to the podcast in headphones and your computer speakers before uploading to the Internet — review, review, review, and find that balance.

To avoid music that's either too soft or too loud, keep the following points in mind as you apply the bed music:

>> **Experiment with levels for the music before you record voice.** Watch your decibel level meter and set your music between –11 and –16 dB, depending on the music or sound effects you're using.

>> **When you're comfortable with the bed's level, lay down a voice track and see how your project's overall levels look (as well as sound) on your decibel level meter.** Remember, your aim is to keep your voice in the 0 or red area without overmodulating. With the music bed now behind your voice, it's much easier to hit the red without effort.

>> **Avoid uneven music, or music that suddenly dips low and then has moments of sudden intensity.** The best music-bed-friendly loops have an even sound (whether driving and dramatic or laid-back and relaxed) and an even level.

>> **Experiment with putting your music at the beginning and at the ending of your podcast instead of throughout.** In these cases, music beds announce upcoming breaks and pauses in your podcast.

Setting volume levels for bed music

Each audio application has its own way to change volume dynamically — but the basic process is the same. In the sections that follow, you find out how to use GarageBand and Audacity to bring music into your podcast at full volume, and then balance it to be just audible enough in the background.

Setting volume levels manually with GarageBand

Follow these steps to set volume levels with GarageBand:

1. **Click the track that your voice resides on.**

All segments in this track only are selected.

2. **Click and drag the playhead to 15 seconds into the project.**

TIP

If you look at the Time Display in GarageBand and do not see a time code, single-click the drop menu at the right of the Display. Doing so accesses the drop menu where you can select the Time mode.

3. **Click and drag the beginning of your vocal track (your voice) to the playhead.**

You now have 15 seconds of lead time between the beginning of the timeline and your podcast.

4. **In the Finder, open a window and find the music file you want to use as your bed music. Click and drag the file into GarageBand, just underneath your podcast track, as shown in Figure 8-5.**

TECHNICAL STUFF

When using sound effects or bed music, try to use AIFF or WAV files. An MP3 file is a *compressed* file. Compressed files (film, photo, audio, whatever) are usually far from ideal for editing because there's loss with every compression cycle — you lose a bit of quality. If you have audio bits you want to export for editing, perhaps in another program, we recommend exporting them in a raw format such as AIFF or WAV so you don't lose any quality.

5. **Select the new track of audio you just created and then choose Mix ⇨ Show Automation to see current volume levels.**

You should see a thin line against your audio tracks. If your audio track is set on the Volume option, the line is your *track volume control*.

6. **Move your playhead to the 10-second mark and click the volume line to create a control point.**

A *control point* is a place you can click and drag to change volume levels at various times. (See Figure 8-6.)

7. **Move your playhead to the 20-second mark and click the volume line to create another control point.**

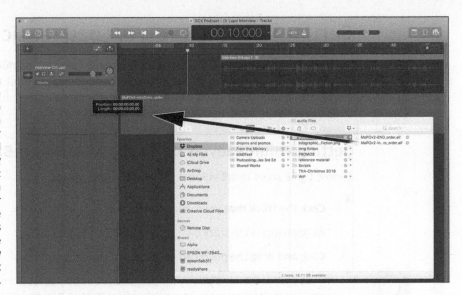

FIGURE 8-5:
A simple drag-and-drop from an open window to your GarageBand window not only imports music but also adds a track to your podcast. Once the new audio is in place, choose Mix ⇨ Show Automation to set levels.

8. **Place your cursor to the right of the second control point you created. Click, hold, and drag the volume line down to as close as -20 dB as you can reach, as seen in Figure 8-6.**

   You have just created a *volume curve.* Now the sound dips lower, allowing for a voiceover to be heard with music softly playing in the background.

9. **Bring the playhead back to the beginning and review your podcast. Change the levels accordingly to set the music at a level you think works best.**

FIGURE 8-6:
Creating a volume curve for a fade-down requires two control points and gives you dynamic control over the track's output.

Keeping volume levels in GarageBand after edits

While adding in bed music, sound effects, or additional audio, you might need to make a change — add or delete another edit. Does this mean you go on and manually adjust your various levels? Not if you confirm Move Track Automations with Regions, located under the Mix menu, is active. This will keep your work up to that point intact while the new audio is positioned in place.

Follow these steps to have GarageBand automatically keep the volume levels in your podcast intact when last-minute audio changes occur:

1. **Find a section in your audio project where you need to add in a segment of audio.**

    Make sure you have levels set in your audio, as shown at the top of Figure 8-7. We will be creating a gap where audio is needed. Make sure you are viewing your Volume Levels, by going to Mix ⇨ Show Automation if you haven't done so in the previous exercise.

FIGURE 8-7:
By activating Move Track Automations with Regions, your audio levels remain intact when major changes occur in your Timeline.

2. **Confirm Move Track Automations with Regions is active by going to the Mix menu.**

   If the Move Track Automations with Regions option is active, a check mark will be to the left of it.

3. **Select segments you want to move to make room for the new audio, then move the selected audio to the right between 5 to 10 seconds (see the bottom of Figure 8-7).**

   The changes you made to the volume levels follow the changes of the audio on the project timeline, also seen in Figure 8-7.

Setting volume levels with Audacity

If you're using Audacity, you can set volume levels by following these steps:

1. **Click the Selection tool to activate it and then click at the ten-second mark of your timeline, placing an edit line there.**

2. **Click the Time Shift Tool button on the toolbar, shown in the margin.**

   The Time Shift tool takes an entire track of content and places it elsewhere along the project's timeline.

3. **Click and drag your audio to the edit line you created in Step 1.**

4. **Choose File ⇨ Import ⇨ Audio and browse for the music file that you want to use as bed music. Then select the file and click Open to begin the import process.**

REMEMBER

   Although you can import MP3 files directly into Audacity when you're editing, we emphasize that you shouldn't be working with compressed files. It's just too easy to lose quality along the way.

5. **Click the Envelope Tool button on the toolbar.**

   The Envelope Tool allows you to dynamically control the audio levels over a timeline. Clicking at the beginning of the audio track establishes the first volume setting, and the second click (at the ten-second mark) establishes the new volume level.

6. **Select the track of recently imported music by clicking its name on the left side of the project window.**

7. **Click at the beginning of the music and then click at the point where your podcast begins (the ten-second mark).**

   Two sets of points appear in your track of music.

8. **Click and drag the second set of points to the 0.5 dB mark.**

   You now have a fade-down of the music that maintains its subdued volume behind the voiceover.

9. **Activate the Selection tool, click at the beginning of the timeline, and then click Play to review your podcast.**

   Now you can change the levels as needed to set the music at a level you think works best.

Making an Entrance: Intros

Now that your podcast has a solid lead-in and a cue to fade out — or you have a loop throughout your podcast that sets a tone — your podcast is beginning to solidify. It's establishing an identity for itself; be proud of the way this podcast of yours is maturing.

But when the microphone comes on, do you always know what those first words are going to be? For some listeners, you're about to make a first impression. What do you want that first impression to be? Are you looking for something spontaneous every time, or do you want to create a familiar greeting that makes listeners feel like old friends? Your chance to make a first impression with your listeners is with an *intro*, which is the first thing the listener hears, be it with a bit of theme music or a word or two about you and your show.

It's up to you, but think about how strong a first impression and a cool intro can make. Consider the ten simple words that became the signature introductions for George Lucas' *Star Wars* saga:

> *A long time ago in a galaxy far, far away . . .*

Or from another science fiction franchise, this time with only four words:

> *Space. The Final Frontier.*

Even if you're not vying to be the next George Lucas or Gene Roddenberry, a good first impression is always important. And let's be honest, something with an original flair and panache is going to make a far more lasting impression than "*Yo, what up? It's your boy/girl,* [insert weblebrity's/influencer's name here]. . ." No matter who you are, this is a moment that can either establish you as a personality (and a podcast) that people will enjoy and eagerly await from episode to episode, or it will make setting you aside from the many other pod people vying for listeners a little tougher. You want the first impression to be unique, lasting . . . and positive.

Consistent, iconic intros can serve as a polished touch of preparation or a subtle flair of professionalism. You're announcing to your audience that the show is on the launchpad, you're ready, and the journey is about to begin.

Theme music

How about a catchy theme? Just as television and motion picture themes establish a thumbprint for themselves in pop culture, an opening theme — be it a favorite song (used *with permission!*) or an original composition — can be just the right intro for your podcast. Perhaps you have a friend in the wide world of podcasting who can assist with audio production. Ask — you might be surprised how willing other podcasters are to assist. Earlier-mentioned software applications such as GarageBand, Cakewalk, Logic Pro, and Audition all offer royalty-free loops that can be easily edited into your own podcast intro. Other great free music sources include Kevin MacLeod's site, Incompetech (`https://incompetech.com/music`), and the YouTube Audio Library (`https://www.youtube.com/audiolibrary/music`), which also includes sound effects. If you have a little cash in the coffers, check out memberships with Digital Juice (`https://www.digitaljuice.com`) for an astounding collection of royalty-free jingles and sound effects.

REMEMBER

Royalty free means you're free to use the audio clip over and over without paying a license fee. You may still need to purchase the audio clips such as Digital Juice. In the case of GarageBand, the loops and clips are included in the price of the package with additional add-on libraries, available on the App Store.

Intro greeting

Some podcasters use quick, snappy intro greetings for their podcasts. For example, Adam Christianson opens every show with a heavy rock riff and the salutation "Hey, Mac Geeks, it's time for the MacCast, the show for Mac Geeks by Mac Geeks . . ." that kicks off his podcast, *The MacCast* (`www.maccast.com`). Others like *Tales from the Archives* (`http://ministryofpeculiaroccurrences.com`) create an imaginative setting as the producers create a more complex intro with music, sound effects, and original dialogue. Whether elaborately produced or just a simple welcome, a consistent greeting serves to bring listeners into your corner of the podosphere.

The elements you're looking for in a spoken introduction are

>> The show's name

>> The name(s) of the host(s)

>> Location of the podcast

>> A tagline that identifies your show

Sit down and brainstorm a few ideas on how to introduce your particular podcast. For example, try these approaches on for size:

>> "Good morning, Planet Earth! You're listening to *My Corner: A Slice of Cyberspace*, and I'm your host, Tee Morris."

>> "From Washington, D.C., welcome to *My Corner: A Slice of Cyberspace*."

>> "With a perspective on politics, technology, and life in general, it's Tee Morris with the *My Corner* podcast."

As you see in these examples, you can mix and match the elements to drop into an intro. Come up with what feels right for you and your podcast and stick with it. After you've put together your greeting, you can either keep it pre-recorded and use it as a *drop-in* (an isolated audio clip that you can use repeatedly, either from podcast to podcast, or within a podcast) at the introduction of your podcast, or you can script it and record it with each session. Whatever method you choose, a greeting is another way of bringing your audience into your 20 to 30 minutes (or 5 minutes to an hour and a half) of time on your preferred media player.

TIP

Regardless of your episode's overall length, try to keep your introduction relatively short. For over a decade, Chuck had a 90-second long intro full of music and audio clips to set the mood, but it was 45 seconds before he would jump in live with "Welcome to Technorama. . .." After some good advice from friends, he shortened the music to 10 seconds to get the audience in to the intro and to the content as quickly as possible. Branding is one thing, but try not to go overboard — 10 to 15 seconds is a good guideline; do some experimenting.

Exit, Stage Left: Outros

Now that you have reached the end of your message or the time limit you have set yourself for a podcast, what do you do? Do you just say "Thanks, everyone, see you next time . . ." or just a basic "Bye," and then it's over? Or do you want to go out with a bit of fanfare? Whatever you decide, an *outro* is much like an intro — as simple or elaborate as you want to make it. Your outro is your final word, closing statement, and grand finale (at least for this episode).

In practical terms, putting together an outro is no different from putting together an intro — same approach, only you're doing it at the end. So review the earlier suggestions for intros and consider what seems a likely direction for an outro.

Some podcasters figure there's little more to think about for an outro than what to say and how to present it. Sure, you could keep it simple — say "Until next time..." and switch off, no fuss, no script, just do it, done, and then upload. But other podcasters see the outro as more of an art form. Before taking a shortcut to the end, check out the following sections for some ways to spiff up your outro.

Leave the audience wanting more

Continuing your podcast to its final moments is a gutsy, confident, and exciting outro, carrying your audience all the way to the final second. This is one of the toughest ways to end a podcast, but if it's done right, it can only make your podcast better.

For example, the *Geek Wolfpack Podcast's ADHD D&D* (http://adhddnd.com) featuring three families and a blind Canadian gamer (https://youtube.com/snowball) connecting on Roll20 (http://www.roll20.net) to play *Dungeons & Dragons* while breaking off into silly tangents ranging from stream of conscious thoughts to various movie quotes. The dungeon crawling usually builds up to a cliffhanger of some description — an upcoming battle, a battle that appears to be turning on the party, or some kind of tension — and then comes an audio cue of a musical sting. The D&D cliffhangers offer a quick and easy way to say "Tune in next week..." without actually doing so.

Catch phrase sign off

Your outro can be the final word from the host, and it should be your bow during the curtain call. A signature farewell is a classy way of saying "This podcast is a wrap. Thanks for listening."

For example, on *Technorama*, Chuck and Kreg have taken a similar approach with their signature signoff starting with Chuck saying "And until next time, a binary high five." To which Kreg always responds "1-0-1." It has become such a well-known piece of the show that listeners writing or calling to leave feedback close their pieces with the same tag.

TIP

If you can't think of anything overly clever, a consistent exit such as "This has been my podcast, protected by a Creative Commons license. Thanks for listening..." works, too.

Credits roll

Another possibility for your outro could be a scripted list of credits: websites where past shows can be downloaded, resources can be endorsed, and special thanks can be given to various supporters of your podcast.

REMEMBER

When listing credits, take care that your list of thank-yous and acknowledgments doesn't ramble on for too long after every podcast. Some podcasters reserve a full list of end credits for special podcasts, such as an end-of-the-season or even final episode. By and large, a minute can serve as a good length for ending credits — plenty of time to mention relevant websites, tuck in the obligatory "Tune in next week . . ." statement, and ask for a vote of support on your favorite podcast directory.

Coming soon to a media player near you

Just as television shows drops teasers of what will be coming up next week, podcasters can also give quick hints as to what is planned for future podcasts. The many podcast fiction titles from Mark Jeffrey (`http://markjeffrey.net`) and Scott Sigler (`http://scottsigler.com`) had already been pre-recorded for the intent of podcasting. This gave both authors a terrific advantage to edit together montages of audio clips and even record a quick synopsis of what will come in future episodes.

Previews for future podcasts tend to be difficult to plan — mainly because of the spontaneous nature of podcasting that the medium prides itself on. Many podcasters have no idea what will be on the agenda for their next show until the day or even a few hours before recording, and then there are other podcasts that start up the audio equipment and speak with no prep time for their latest installment. However, for those podcasts that can provide glimpses of things to come, this kind of outro serves as a commitment to the audience that there will be more content coming through the RSS feed and that programming is being planned for future installments.

As mentioned with intros, your outro can use one of these approaches or combine them. Find what best fits your podcast and stick with it. The more consistency your podcast can follow, the more professional it sounds — spontaneous but focused, right?

3

So You've Got This Great Recording of Your Voice: Now What?

Chapter **9**

Shrink That Puppy and Slap a Label on It

You've finished your final edit (and if you think we're skirting blasphemy whenever we use the *E* word, trust us — creativity sometimes demands a sacrifice). At last you're ready to compress your mondo-super-sized AIFF or WAV file down to the format that invokes traumatic memories from RIAA representatives everywhere: *MP3*.

The MP3 format was designed to reduce the amount of data (via compression) required for digitized audio while still retaining the quality of the original recording. MP3 files are the best way to keep the audio small enough in size to make it a quick-and-easy download, and it's this format that podcasting uses to get content efficiently from podcaster to podcatching client. Although creating MP3s is a simple enough process, you do need to make some tradeoffs between quality and compression.

A Kilobit of Me, and a Whole Lot of You: Understanding Kbps

The compression process begins with proper *bitrate* settings, measured in Kbps (kilobits per second). Bitrate is a method of measuring data transmission from one point to the next. The higher the Kbps value, the more data being transferred between two points. The more data being transferred per second, the better the quality of information. With each rate of data transfer offered by recording applications, you can digitally reproduce the various qualities of audio:

>> **8 Kbps,** matching the vocal quality of a telephone conversation

>> **32 Kbps,** yielding audio quality similar to AM radio

>> **96 Kbps,** yielding audio quality similar to FM radio

>> **128 Kbps,** matching audio CD quality, and the most common Kbps used for MP3 compression of music

>> **192 Kbps,** a "Higher Quality" setting Kbps used for MP3 compression of music; better audio sound, but larger data files

You aren't married to this listing of Kbps by any stretch of imagination. Tweak 'til you find the Kbps best suited for your podcast.

REMEMBER

It's a tradeoff: The higher the bitrate, the larger your file size is; but the smaller your bitrate, the lower your audio quality is. A good idea is to experiment. Compress your podcast using one bitrate, save it that way, and then change your bitrate to something higher (or lower), save it, and play back each clip to compare how they load and how they sound. Note any changes in sound quality and file size. When you find that happy medium between quality and compression, stick with that number for your current and future podcasts. If you want to skip experimentation, refer to the suggested compression settings we have outlined on the Cheat Sheet available online.

As you can see throughout *Podcasting For Dummies*, we're leaning heavily on Audacity as your audio solution. There's a good reason for that: It's a rock of reliability. While it is not a perfect application, guess what? No application is perfect. Audacity has been around since the beginning of podcasting and has only gotten better over time. In this chapter, we go deeper into what you can do with audio on a more technical level.

Changing bitrates in Audacity was once a multistep process, but with recent releases, creating MP3 files has become incredibly easy:

1. **Import your audio (a WAV or an AIFF of your edited and polished audio file) into Audacity.**

2. **Choose File ➪ Export ➪ Export as MP3.**

3. **Select from the Format Options section (see Figure 9-1) one of the following Quality settings:**

   - Medium (145–185 Kbps)
   - Standard (170–210 Kbps)
   - Extreme (220–260 Kbps)
   - Insane (320 Kbps)

   You can use a setting higher than the Standard setting but that will only make a difference if you are using higher-end (professional grade) headphones for listening.

WARNING

Saving at a higher quality like Extreme or Insane also means any faults or hiccups in your audio are preserved. If you are editing and reviewing your audio on pro-level equipment, you are more than okay exporting your podcast's MP3 at the Standard level.

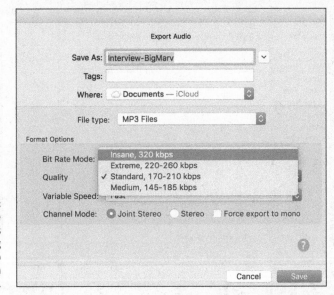

FIGURE 9-1:
You find a wide variety of bitrates for compressing your podcast to MP3 with Audacity.

4. **Confirm that your Variable Speed is set at Fast, and your Channel mode is set to Joint Stereo.**

   These settings are the default when you initially install and launch Audacity. These settings remain the best for your audio.

   Bit Rate Mode (Constant) and Channel Mode (Joint Stereo) are two final details you should take a moment to set up. Constant Bitrate ensures the bits are output at a steady (constant) rate despite the audio that is being used (including silence). Variable Bitrate attempts to do a bit of compression — for example, the bitrate of a complex music sample would take more space than that of the same length of time in a spoken piece with several pauses. Although variable length audio sounds good on the surface, it can present problems to the listener — actual playback times may not be correct, causing problems when people pause a long podcast. *Joint Stereo* refers to a technique that saves some space. For more information, refer to the sidebar "Rocking the podcast joint with Joint Stereo."

5. **In the Save As field, name your file.**

6. **Click the Save button.**

7. **When prompted, fill out your ID3 Tags (as seen in Figure 9-2).**

   For more on ID3 Tags, see our ID3 Tag section at the end of this chapter.

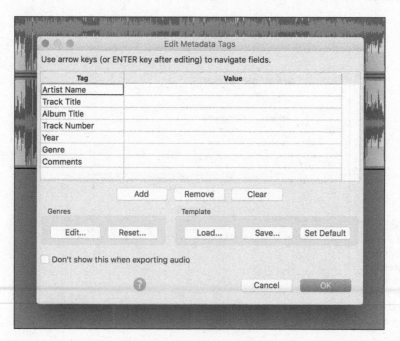

FIGURE 9-2: Fill out your ID3 Tags.

ROCKING THE PODCAST JOINT WITH JOINT STEREO

Let's clear the air about stereo and joint stereo:

- **Stereo:** By default, many applications that export to MP3 also export as Stereo. Stereo encoding takes the left and right audio separately and divides the selected bitrate for each. The drawback to (plain) stereo is that the channels are equal width, so a 128 Kbps encoding of a mono signal is really two identical 64 Kbps channels, resulting in a 50 percent inefficiency.

- **Joint Stereo:** With this option, one channel carries sound that is identical in both the left and right stereo tracks, and the other channel carries the difference. Both channels are only as big as they need to be, so if your podcast is mostly mono — with just one microphone — you'll get more out of your bitrate than you would with a plain stereo file, but without the reliability issues of a mono file.

When working with a consistent sound for your podcast, working in one stereo output is something you should consider. Setting standards for yourself and your podcasts lays solid groundwork for success. For the writers, their own experiences have them using Joint Stereo as the go-to podcast format.

After you set the bitrate for this one file, it's one-and-done — your settings are automatically applied to other MP3 files created from this point. Only when you go into the Options for MP3 files will you change bitrates.

Care for a Sample, Sir? (Audio Sample Rates)

When you have a grasp of bitrates, you're ready to move on to *sample rates*. This may strike you as a tad redundant, particularly when you see the list of common audio-sample rates. A strange *déjà vu* makes you wonder whether you're in the real world or exist merely as part of The Matrix. No biggie — sometimes technology *is* redundant.

As we discuss in the previous section, bitrate is a measurement of how much audio data is transferred between two points, such as the computer and your headset. The *sampling rate* determines the maximum sound frequency that can be reproduced — the value you set is twice the frequency value. For example, a

44.1 kHz sample rate can reproduce sounds up to 22.05 kHz, which is slightly above the range of human hearing.

Here are a number of audio sampling rates found in MP3 encoders:

>> **8,000 Hz/8 kHz,** matching the audio quality of a telephone conversation

>> **22,050 Hz/22 kHz,** matching audio quality similar to AM radio

>> **32,000 Hz/32 kHz,** matching audio quality similar to FM radio

>> **44,100 Hz/44 kHz,** matching audio CD quality

>> **48,000 Hz/48 kHz,** matching digital TV and film audio, and maximum sound quality on professional-grade audio/video recording

As with your audio-recording applications, you have a range of sampling rates available — and you can type in your own custom sampling rates.

REMEMBER

Using different sample rates in compressed audio files does not affect the file size — only the bitrate and duration affect file size. A high sample rate at a low bitrate can result in poor signal representation and thus poor quality, but if you're using at least 64 Kbps (bitrate), there's no good reason to ever use anything except 44.1 kHz as the sample rate.

WARNING

Be careful when using a sample rate other than a multiple of 11. Choosing a sample rate other than 11.025 kHz, 22.5 kHz, or 44.1 kHz can produce an effect similar to the voices of Alvin and the Chipmunks. It might be funny at first, but most listeners won't stick around to hear your entire message.

Sample rates are located in the Preferences window and it is extremely simple to either change or customize to your podcast's personal needs. Follow these steps to customize sample rates in Audacity:

1. **Choose Audacity ⇨ Preferences (Mac OS) or Edit ⇨ Preferences (Windows).**

   The Preferences window appears.

2. **Click the Quality option in the left column.**

3. **Select your desired bitrate from the Default Sample Rate drop-down menu (as shown in Figure 9-3) or enter your own custom sampling rate by selecting the Other option and entering a number.**

REMEMBER

If you're entering a custom sampling rate, make sure you enter the rate in *hertz* (Hz) and not *kilohertz* (kHz). For example, if you decide your sampling rate is 44.1 kHz, you have to enter **44100**, not 44.

4. **Click OK.**

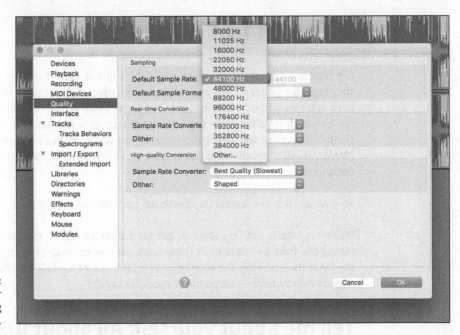

FIGURE 9-3:
Changing your
audio sampling
rate in Audacity.

Is it soup yet? Not quite yet. You may have heard us talk about ID3 tags, but what exactly are they? That's the final step before you upload the podcast to your website, and some podcasters — and we're talking experienced podcasters from the early days — still refuse to tend to this detail.

While the rebels in us really admire their digging in the heels and not conforming to the rules of ID3 tags, we as long-time podcasters have something to say to you who defiantly refuse to tend to this final detail: Enjoy netcasting, you nimrods.

And thank you for buying our book and attending our TED Talk.

Yes, ID3 tags matter. This batch of details tells other podcasters who you are, tells listeners what your podcast is all about, and gives a quick overview for which episode they're listening to at the moment. ID3 Tags do a lot to give your podcast a touch of flair, a hint of polish, and not only include details on your episode, but showcase the artwork unique to your podcast.

The next section takes a look at exactly what they are.

ID3 Tags: The 411 of Podcasting

ID3 Tags are a final detail, some podcasters either skip it or just don't care how they tag their podcasts. As you can see in the preceding section, we take exception with extreme prejudice about that. ID3 tags, when ignored, obliterates any hope for listeners to effectively organize podcasts in a computer's media player or remind drivers of what they are listening to while on their commute. The bigger platforms, like Spotify and Apple Podcasts, are now *insisting* that ID3 tags are completed.

So yeah . . . it's not just us OPs (original podcasters) saying this.

While we totally get you want to get your content out to the fans ASAP without taking the time to implement these tags, we entreat you: Stop the madness, stop the insanity, and stop the monkeys. Proclaim your true self for the sake of media players everywhere — implement your ID3 tags!

Tell me about yourself: All about ID3 tags

First created in 1996, ID3 tags were designed to be added to audio files in order to have the artist, album, and track title displayed in a computer's MP3 player when the file is played back. Current ID3 tags now include composer, bitrate, album art, and even genre.

Of course, another question that comes to mind is *why* would a podcaster want to even bother with ID3 tags?

Take a look at a podcast you're listening to, regardless of whether you are old schooling it on a portable MP3 player or listening through your desktop computer. Your podcast app probably organizes your various feeds by the date downloaded, either with the most recent show at the bottom or at the top of your playlist, depending on the player's preferences. Each individual podcast has an episode title, an artist, and a podcast (show) title. If the podcaster working with the ID3 tags is particularly savvy, artwork associated with your show is also displayed.

Now hop from podcast to podcast in your MP3 player, and you can tell which people care — or don't — about identifying their shows. Some podcasts simply use obscure numbers that could be a date, but when you hear two episodes of the podcast back to back, you find out the number and the date read at the intro of the show (provided there is one) don't match up (confusing). With ID3 tags in place, you can now look at your player and get an idea of exactly what you're listening to.

When you add ID3 tags to your podcast, you set apart individual podcasts from one another, making each one unique but keeping it grouped with your podcast show.

IDentity crisis: Making ID3 tags work for podcasting

Some ID3 tags really don't work for podcasting. Album? Track number? Composer? You're podcasting, not producing music. Your responsibility (as a podcaster) is to redefine the following tags we have found useful for the podcasting medium:

>> **Artist Name:** This one is pretty self-explanatory. You're the artist behind this podcast, so let people know who you and your group are. Put in your name (or at least the pseudonym you're using as a podcaster) or the name of your podcasting team. For example, `Philippa Ballantine`, `August Grappin`, and `D.J. Steve Boyett`.

>> **Track Title:** This should be the name or number of your episode. Examples are `Show #19`, `Episode 3: One of Those Faces`, `BOOK THREE: Chapters 39 & 40`.

>> **Album Title:** Here is the name of your show, your show's website, or your network if you are part of a podcast network. Good examples are `The Shared Desk`, `Destiny Community Podcast`, and `Podrunner`.

>> **Album Artist:** Many times left blank, this tag can be a bit tricky for users. You can use it to refer to the show artwork artist, the production studio, or the creator of the podcast. Examples are `Imagine That! Studios` and `Time Magazine`.

>> **Track Number:** Purely optional, this ID3 tag allows you to make sure your podcasts remain in some kind of sequential order. For podcast novelists, the track numbers coincide with the chapter numbers. If podcasts follow a season of multiple episodes, the Track Number coincides with the episode in that season.

>> **Year:** The year this podcast episode was produced.

>> **Genre:** The genre Podcast wasn't offered in drop-down menus of MP3 creators, but with the growing popularity of the medium, it's becoming more and more common.

TIP

When using Audacity, if you don't see the Podcast genre, you can manually type **Podcast** into the field.

>> **Comments:** Similar to comments you leave in XML, you can give a quick two or three lines of show notes for your podcasts. This field is a great place to put in any Creative Commons notices, websites for more information of the show and its hosts, and special dedications.

If you are using another app for your ID3 tags, you may also be asked to complete additional tags, including:

» **Composer:** This tag can be used to feature the name of your producer or head editor. You can enter your engineer's name, or the studio where the show is produced. For example, *The Shared Desk* artists is listed as `Imagine That! Studios` but the composers are `Tee Morris & Pip Ballantine` as the show is co-produced and co-hosted.

» **Grouping:** This tag may remain blank until you're affiliated with a network or distribution hub of some kind. A good example would be `QuickAndDirtyTips.com` where podcasting superstar Mignon Fogarty leads a team of podcasters all offering sound advice and quick life hacks.

» **BPM:** If you are into staying fit and are producing a workout podcast, you want to fill in this ID3 tag as BPM is for Beats Per Minute. BPM is also good if your podcast is a house music-dance mix podcast. A good example of a podcast showing off BPM is any one mixed by D.J. Steve Boyett (`https://podrunner.com`).

Reminiscent of John Cleese's "Adapt, and improve. That's the motto of the Round Table" aside in the *Monty Python* robbery sketch; we podcasters must adapt these ID3 tags to our podcasts to improve how they appear in our players. On playback, the ID3 tags appear on the listeners' interfaces, offering a quick glance at the content of the podcast. Figure 9-4 shows how this blast from the past appears on Tee's dashboard.

FIGURE 9-4:
How a properly tagged podcast appears on a car's stereo system.

WARNING

A peril in working with ID3 Tags is spelling errors. There is no spell check in Audacity or any other tag editor. Because many directories, search engines, and players rely on the ID3 tags to organize and properly display your show, make sure to double-check the spelling of your show title, names of people involved, and any other text you're putting in place. Also, some editors will create quick fill databases that help you fill in your File Info quickly. This database also includes the misspellings. When filling in your ID3 tags, take your time and check your spelling.

Creating and editing ID3 tags

Audacity lets you set up a template for the ID3 tags that gets applied when MP3 files are created. To create this template, open Audacity, import a file or a project that is ready for ID3 compression, and follow these steps:

1. **Choose Edit ⇨ Metadata from Audacity's main menu.**

The Edit Metadata Tags (ID3 Tags) window opens (see Figure 9-5).

TIP

You can also create ID3 tags and ID3 tag templates by choosing File ⇨ Export ⇨ Export as MP3. After setting Bit Rate, Quality, and other technical details, click the Save button. The ID3 tag window appears automatically.

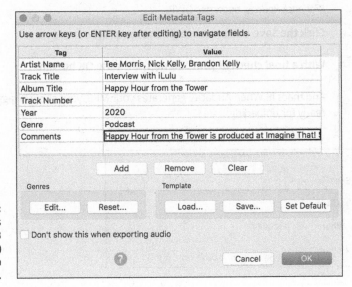

FIGURE 9-5:
The Edit ID3 Tags (for MP3 Exporting) window in Audacity.

2. **Fill in these fields:**

- **Artist Name:** The podcaster or name of the podcasting crew.

- **Track Title:** The episode name and/or number.

- **Album Title:** The show's name or website.

- **Track Number:** The field is optional, unless numerical order is a priority in your podcast.

- **Year:** The podcast publication's year.

- **Genre:** Type Podcast into this field.

- **Comments:** Type or paste into this field the following: [TITLE OF YOUR PODCAST] is produced by [PRODUCER OR PRODUCTION STUDIO] and is protected by a Creative Commons 3.0 license. You can find out more about this license at creativecommons.org and visit our website at [PODCAST WEBSITE] for more information.

3. **After proofreading your ID3 tags, go to the Template section of the Metadata Tags window and click the Save button.**

4. **Name for your ID3 tags (see Figure 9-6).**

   Make the name of your template an acronym from your podcast's full title. For example, the *Happy Hour from the Tower* podcast is HHFTT and the *Geek Wolfpack Podcast* becomes GWP.

5. **Click the Save button to create the template.**

6. **With a final check of your ID3 tags, click the OK button.**

   And that's it! When Audacity generates your MP3, your show is tagged and ready for uploading.

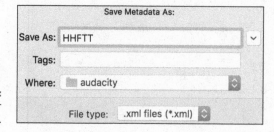

FIGURE 9-6:
Name your
ID3 tags.

And when you get to your next project . . .

1. **After you finish editing and saving a new project, choose Edit ➪ Metadata from Audacity's main menu.**

2. **Go to the Template section of the Metadata Tags window and click the Load button.**

3. **Find the XML file with the name of the podcast you are working on and click it.**

4. **Click the Open button to create the template.**

5. **Go to Track Title and change the title to reflect the subject matter of the current episode.**

6. **With a final check of your ID3 tags, click the OK button.**

After you tag everything properly, there's one more piece of identity to apply to your spectacular creation — artwork. We cover artwork in Chapter 10 because it's one of the files you'll need to move to your server.

AUTOMATING WITH AUPHONIC

Face it: Post-production is one of the least favorite parts about podcasting. The temptation after you stop recording is to upload and move on with your life, but there's editing, tagging, uploading, ugh! One of the best ways to keep podcasting from becoming a chore is to automate wherever possible. Enter Auphonic (https://auphonic.com). This online service takes your audio or video files and automates the post-production with templates to reduce the time it takes to do all the chores, including tagging, putting on an intro, outro, and artwork. It can even upload your final production to your server, Dropbox, LibSyn, or Amazon S3 server (see Chapter 10). The free service offers two hours of processing per month. Depending on your show length and frequency, you may find it worthwhile to invest a little bit to save yourself some time.

Auphonic supports a mobile app to allow you to literally press record, talk, stop, and have it do its thing. All you need to do is write the blog entry to get your RSS feed updated. Chuck uses this as his mobile rig quite a bit when he's on the road doing his slice-of-life podcast.

Chapter **10**

Move It on Up (to Your Web Server)

You've managed to figure out what it is you want to say (or show), you've gone through the trials and tribulations of the editing process (or not), and you've faithfully employed correct ID3 tagging (non-negotiable). That's great, but no one is going to hear your contribution to the podcasting world until you put your files up on the web.

In Chapter 1, we cover the hosting provider selection process. In this chapter, we take an extensive look at the mechanics of the process, including how to appropriately name and organize your files.

Podcasters have a variety of options when it comes to uploading files. Although the methods are all different, they all help you accomplish the same job: copying files from your personal computer to their new home online. Many hosting providers have easy-to-use browser-based drag-and-drop file transfer utilities built into their service. In many cases, this may be all you need to get your files onto your server for others to consume. In this chapter, we dig a little deeper to ensure you not only know the "Ya' Ba-Sic" way of doing things but that you understand other options to make it easier in cases where you need to maintain files for multiple shows or multiple hosting providers.

Show Art: Getting Graphic with Your Podcast

A big piece of successful podcasting is marketing, and a big part of marketing is *branding*. Your podcast logo (or *show art*) is very important to help set you apart. A large percentage of listeners use apps where show art is prominently displayed, both in directories and in playback modes. Apple once was indifferent about show art, but now it's part of registering your podcast with its directory. So when getting ready to launch your podcast, you need to have a good look for your production like those featured in Figure 10-1.

FIGURE 10-1: *Show Art* is a podcaster's way to brand a show and give your episode a slick final touch when played back on various media devices.

Show artwork, commonly seen in many media players, is a nice option for podcasters who want to brand a podcast with a logo. Mur Lafferty's Parsec-winning *I Should Be Writing* logo is a classic broadcast microphone with a sticky note slapped on it and a pencil, sharpened and ready for use. Then you have *The Onion Radio News*'s trademark onion with a globe ghosted behind it, the stamp of quality journalism at its funniest. These icons are associated with their shows, and this kind of branding is becoming more and more common in podcasting.

To get your artwork ready for prime time:

1. **Design your show art to the following specifications:**

   - **Format:** JPEG (.jpg) or PNG (.png) format

   - **Color Mode:** 8-bit channel, RGB mode

   - **Resolution:** 72 dpi

   - **Size:** 1400 × 1400 pixels (minimum)

     or 3000 × 3000 pixels (maximum)

2. **Using File Transfer Protocol (FTP) application, upload your art somewhere on to your web server.**

   If you are using a service like LibSyn, you will use its FTP options the same way you would upload an episode. If you are hosting the podcast on your own, you can use Cyberduck or Fetch to upload it somewhere on your server. For more details on uploading files to a server, take a look at the "Uploading your files" section, later in this chapter.

3. **Pull up the artwork on your browser and then copy its URL.**

   Once you have the URL of the show art copied, have it on a note somewhere on your computer. You will need that URL when registering your show with various directories.

4. **When using your podcast plug-in with your blog, enter in your artwork's URL when asked for it, as seen in Figure 10-2.**

TIP

   If you do a makeover of your show art, you can upload a new image to replace the old one.

5. **Save your changes in the plug-in.**

After you have all the details of your file covered, down to the branding behind your podcast, you have to get everything online so that others can enjoy what your creativity has to offer.

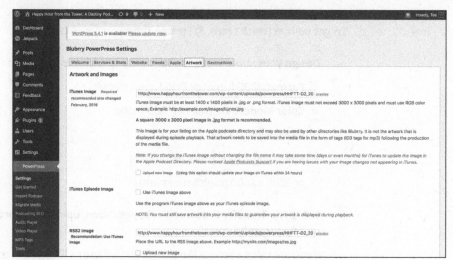

Adopting an Effective File Naming Convention

In Chapter 9, we talk about the importance of "the little things" such as ID3 tags. Equally important is how you decide to name your podcast media files. In this section, we illustrate the importance, not only to you as a podcaster, but also to your listening audience. Although no hard and fast rules for naming files exist, following some common conventions allows everyone to easily find your podcasts.

A good naming convention of a podcast accomplishes the following:

>> **Easy sequential ordering:** Files should not appear at random in your directory. They should line up — first, second, third, and so on.

>> **"At a glance" recognition for your listeners:** Calling a podcast media file my media file doesn't help very much. Calling it Bob's Fencing Podcast certainly does.

Here's an example of a well-named podcast media file, if we do say so ourselves:

```
Tech_Ep600_20200516.mp3
```

Although the structure may not be obvious, this filename adheres to both rules and even adds one more:

» **Tech:** This abbreviation stands for *Technorama*. Starting the filename with these four letters organizes the files together in the media folder. For listeners who can see only the filename on their MP3 players or computers, they can quickly recognize that files starting with these four letters belong to his show. Pretty much everyone listens to podcasts through some podcatching client these days, but don't discount that small segment that likes to download files to their hard drive and listen.

» **Ep600:** This is Episode 600 of the show. Referring to each show with a sequential number is a good idea, giving you and your fans a common reference point that is easier than "Remember when you did that one show with that one guy who said that funny thing? Man that was great!" Having the episode number right after the name also ensures that the files "stack up" right in the media folder, as well as in your listeners' MP3 players. You may want to consider adding some leading zeroes to the number so episode 3 (003) comes before 600. Add more zeroes if you plan on doing more than 999 shows.

» **20200516:** The date Chuck posted the file. The order he uses here is year-month-day. He uses 05 instead of 5 so that dates from October (10), November (11), and December (12) don't intermingle with January (which would happen if he used 1 instead of 01). True, in this case the files were already sorted by episode number, but you may choose to go with the date first.

Between each element, he adds an underscore (_) simply to provide a clear distinction of each part for his eye, or anyone else's eye looking at the filename.

WARNING

Don't use any spaces or special characters in your filenames. Stick with A–Z, 0–9, dashes, and underscores. Slashes (/, \), octothorpes (#), ampersands (&), and others can and do cause problems when creating RSS 2.0 files or when clients are handling the file. Also, leaving a space in the filename might mess up a URL, so if you want to space things out, use underscores or dashes (for example, use My_Podcast.mp3 instead of My Podcast.mp3).

The following examples are from well-respected podcasters who all follow the guidelines we set forth earlier:

» NMS-2017-04-01.mp3

» scienceontriplej20181006.mp3

» AC_EPISODE_021.mp3

» TSD-075.mp3

Note how each of them identifies the name of the podcast and provides a sequential way of ordering the files. If you leave out the episode number, as in the case of the Science on Triple J file, the year of the podcast ahead of month and day also ensures that 2018 files are always grouped together. If the podcaster had used the month first, as people traditionally think of dates in the United States, files would be mixed based on the month they were released, regardless of the year.

If thinking about the date that way seems a little too strange for you, do what Chuck does — stick a sequential episode number in front of your date, and don't worry about it.

Understanding How FTP Works

FTP (File Transfer Protocol) is the method by which you can transfer files from one destination to another over the Internet. You likely transfer files every day, from your desktop to your documents folder, from an email to your desktop, or even media files from a podcast's server to your podcast app of choice.

Transferring your podcast files to and from the Internet isn't much different, at least on the surface and even at any depth necessary for podcasters to ply their trade. Lucky for you, specialty software exists to make this process even more simple. FTPing files has become as simple as dragging and dropping.

TIP

Some podcast hosting services, such as *Liberated Syndication*, or LibSyn (`http://libsyn.com`), mentioned earlier in the Show Art section, make this simple process even simpler by providing a web form to handle the uploading of podcast files, as described later in the "Uploading to a Podcast-Specific Host" section. Browser-based systems certainly remove the complexity for many, but it doesn't hurt to understand the processes outlined in this chapter. Many experienced podcasters need more flexibility than the limited functionality a browser-based upload process allows.

Regardless of what software, forms, or other assistance you use to move files around, the concept of FTP is the same.

FTP has been around for quite a while now. Archaic and seemingly nonintuitive names abound from the start, such as the following two computer systems involved:

>> **Local host:** The local host is the computer you are sitting in front of and initiating the file transfer from. If you're using a laptop to connect to your web server, your laptop is the local host. If you're at work and logged in from a workstation, your office computer is the local host.

The *local directory* or *local path* is the folder on your local host that contains the files you want to transfer. You can change local directories at will, but most FTP programs have a default starting place. Feel free to move around after that.

>> **Remote host:** The remote host is the computer or web server to which you've connected. It's likely the spot where you're trying to get your MP3 files to go to allow others to download them.

Not surprisingly, the remote host has its own *remote directory* (the folder on the remote system where you drop your files). Again, you can change or navigate through remote directories just as you can change the file folders on your computer.

Making Your Connection with an FTP Application

You need three pieces of information to initiate an FTP connection:

>> The IP address or hostname of your remote host

>> Login name

>> Password

Your hosting company should have provided this information to you. If you don't have it handy, find it. You're not going any further without it.

All FTP programs do the same job but have slightly different methods of going about it. After you grasp the concept, using just about any FTP client is a simple process. Here are the general steps you follow to set up a connection in any FTP client:

1. **Launch your FTP client and create a new connection.**

   Because this step is what FTP clients are designed to do, they usually make this process very simple.

2. **Enter the hostname of your web server, username, and password.**

   This step identifies the remote system and shows you have access to the files and folders it contains.

3. **Connect using either a button or a menu option.**

   Depending on the speed of your connection, the connection is established in a matter of seconds.

The following sections show you how to use Cyberduck (http://cyberduck.io) and FileZilla (http://filezilla-project.org). Both apps are free and available for Windows and Mac. You can find many FTP programs as freeware, shareware, and shrink-wrapped software, for every brand of up-to-date operating system. We picked these two for their ease of use and streamlined approach to getting the job done, but you can use the FTP program of your choice.

Step by step (or quack by quack) setup for Cyberduck

After you download the Cyberduck program from http://cyberduck.io onto your Mac, you can follow these steps to set up an FTP connection:

1. **Click the Open Connection button in the upper-left corner.**

   The Connect dialog box appears, as shown in Figure 10-3.

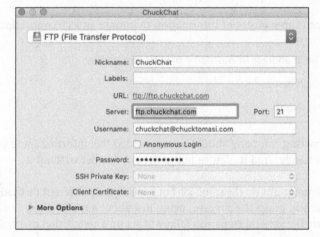

FIGURE 10-3:
A properly configured Cyberduck FTP connection.

2. **In the Server text box, enter the name of your server.**

   Depending on the requirements of your ISP, this name can be in the format of *ftp.mydomainname.com* or perhaps simply *my_domain_name.com*. And of course, you need to be sure and use the name or IP address of your web server. Chances are good that you don't really own the domain *my_domain_name.com*, right?

3. **Enter your Username and Password in the text boxes.**

The hosting company should have provided this information. If your hosting company is the same company that is supplying your connection to the Internet, it might be the same information you use to check your email. But if you toss some additional money each week at a hosting provider, it's likely something completely different.

For Mac users, select the Add to Keychain option to store your username and password for the next time you connect.

4. **Click the Connect button in the lower-right corner.**

If you entered things properly, you now see the file folders on your remote web server. If you didn't, you get an error message or a login failed dialog box. Correct what's wrong and try it again.

When your connection is established, choose Bookmarks⇨ New Bookmark. Give your newly created connection a catchy name and simply double-click the given name the next time you need to connect.

Step by step setup for FileZilla

FileZilla (http://filezilla-project.org) operates much the same as Windows Explorer or the Mac Finder, allowing you to drag and drop files between your PC and your FTP server. When you've downloaded the FileZilla program, follow these steps:

1. **On the Login toolbar (shown in Figure 10-4), enter your server name, username, and password.**

Leave the port blank. (It defaults to 21.)

2. **Press the Quickconnect button.**

If you did things right, you see a lot of text scroll by in the upper-most window and content from your web server appears in the right windows, labeled Remote Site.

Congratulations; you're now connected to your web server. If you entered something wrong, you get an error message in red text in the upper window. Correct your mistakes and try it again.

From here, you can navigate through the folders on your web server much as you do on your computer's hard drive. You can move up or down the file system, finding the spot where you want to drop your podcast files.

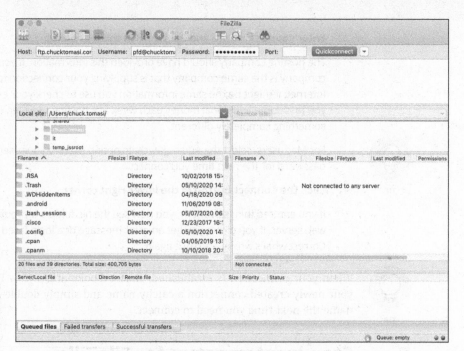

FIGURE 10-4:
A properly configured FileZilla connection.

A place on your web server for your stuff

Logging in to your web server for the first time can be an intimidating process. In this section, we show you how to place your files in a location so you can easily create links to your podcast files.

Don't be intimidated by the odd directory names on your web server. The only one you really need to know is `Public_html`. Other hosts may call it `www` or simply `html`. If you don't have one of those three, poke around until you find one that has a bunch of files in it that end in `.html`.

WARNING

Look, but don't touch. Going into folders doesn't hurt anything, but doing silly things such as deleting, renaming, and moving files you know nothing about is a bad idea. All those strangely named folders do something, and they're likely necessary to make your website work right. Remember the proverb "'Tis best to leave functioning web servers lie. . . ."

You may see lots of different files and a few folders. We show you how to add even more files to this system, so now is a good time to think about organizing and housekeeping.

Start by creating a special place to keep your podcast files. Making a new folder exclusively for your podcast media files not only separates your podcasts from your other critical web server files, but it also allows you to quickly see what is currently live and what needs to be cleaned up.

In your root directory (the top-level folder usually denoted with a forward slash), create a new folder called media. With Cyberduck, choose File ⇨ New Folder, followed by entering the name in the resulting dialog box. On FileZilla, right-click the window with the details (date, size, permissions) on the remote file server and choose Create Directory.

After you create the new media folder, double-click the name of the folder to open it. You're now inside your totally empty media folder, and ready to load it with your podcast media files.

Uploading your files

After you set up a folder for your podcast media files and decide on a file naming convention, you're ready to move your freshly named files to the web server.

Both Cyberduck and FileZilla support *drag-and-drop* file transfers. If you're new to FTP, the FTP program interface may be easier for you to use.

For Cyberduck, follow these steps:

1. **Choose File ⇨ Upload.**

2. **Browse your system to find the podcast media file you want to upload.**

3. **Click the Upload button.**

For FileZilla, here's what you do:

1. **Navigate to the desired folder using the Local Site window on the left.**

2. **Select one or more files and/or folders from the window directly below.**

3. **Drag the selected files to the Remote Site window on the right.**

Depending on the size of your file and the speed of your Internet connection, the file may transfer in a matter of seconds or a matter of minutes. When completed, the file appears in your FTP client and is ready to be linked in your show notes and RSS 2.0 feed.

USING YOUR BLOGGING SOFTWARE TO UPLOAD

Many popular blogging tools allow podcasters to upload files without the need for special FTP software, much like the HTTP process.

However, podcast media files often exceed the file size requirements for these services. For this reason, we recommend not using your conventional blogging software to handle your podcast media file uploads.

Consider that blogs are primarily used to communicate text. Although you can easily extend the functionality of a blog to support a podcast, the site management tools are designed for text and images.

WordPress (http://wordpress.org) does allow file uploads, but the file size depends on system parameters that are often set by the web host and can't be changed by the individual. You can upload, but it does take a significant amount of work that exceeds the scope of this book.

Instead, we recommend using blogging software to manage the posting of your content after your file is uploaded to the server. In WordPress, this is easily done with a plug-in called PowerPress (https://create.blubrry.com/resources/powerpress/).

Uploading to a Podcast-Specific Host

Podcast-specific hosting companies significantly simplify the uploading process; many include web-based forms that take the place of additional computer programs to handle the uploading process. They also take care of archiving, RSS 2.0 creation, and even ID3 tagging.

TIP

Although this web-based uploading process is simple, we prefer the flexibility of using an FTP client — or better yet a command line interface. Or maybe we're old school. . . .

For the purposes of illustration, we use an account with LibSyn in this section. If you haven't already, you need to sign up with LibSyn and create your own account. Then follow these steps:

1. **On LibSyn's home page, enter your username and password and click the Login button.**

2. **Click the Content menu option and select Add New Episode.**

   You are taken to the New Content page, as shown in Figure 10-5.

3. **Click the Add Media File button.**

4. **Click Upload File From Hard Drive if your files are on your local hard drive. There are other options available if your file is on another server accessible via FTP or Dropbox.**

   The File Upload dialog box opens.

   Note: *HTTP upload* is another name for form-based transfers.

5. **Find the podcast media file you want to upload and either double-click the file or select it and click Open.**

TIP

If you've used proper ID3 tags on your podcast — and you have by this point, right? — you can check the box "Populate Form with ID3 Data" to save some time. Just be aware that the description may need some additional formatting as the importer doesn't recognize paragraph breaks. Also check your title and description for "encoded" characters. You may need to do touch-ups like replace & with the & character.

6. **If you did not choose to import the content of the ID3 tags, enter the information for your blog and/or podcast in the form.**

   The blog settings are very simple: things like the name of your blog, your email address, and what category you want it placed in. Nothing here is mission critical, so fill it out however you want to see it listed. You can always come back and change it later.

7. **When you finish, click the Publish button.**

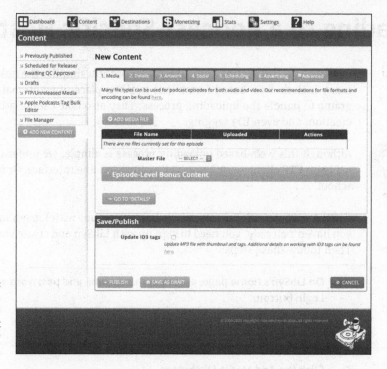

FIGURE 10-5:
Adding media files to a LibSyn account doesn't require an FTP client.

While LibSyn has full hosting capabilities, some may find the blogging feature a bit limited and use LibSyn as a file repository while hosting their blog on their own site, then referencing the files on LibSyn from their blog. To access files stored on LibSyn in your blog (and resulting RSS feed) follow the preceding steps to upload and publish the content. Then follow these steps after your file is published.

1. **Go to the Content menu and select Previously Published.**

2. **Locate the file in the list and click the Link/Embed icon for that episode.**

3. **Copy the URL from the Direct Download URL field and use that as the location of your file in your blog software as needed.**

TIP

LibSyn offers an FTP interface if you prefer to use an FTP client or command line interface. Your login and password are the same as through the web interface.

TIP

If you like to try it before you buy it, LibSyn offers up to a free month with the promo code *podcast411* allowing you to get a feel for how it works before you fork over your hard-earned cash. Not that the very inexpensive monthly fees will break you or anything, but it's nice to know how your future home might work.

IN THIS CHAPTER

» **Understanding good show note etiquette**

» **Planning your show notes**

» **Deciding on your level of detail**

» **Using images effectively**

» **Posting with searchers in mind**

Chapter **11**

Providing Show Notes

S how notes are brief summaries of each podcast episode. Show notes can take the form of an outline, a detailed bulleted list, or just a few sentences of text. In this chapter, we show you how to effectively use show notes to enhance the listener experience of your show and bring in additional traffic to your podcast through search engines.

And where do you find these show notes? Simply enough, on the podcast's website.

REMEMBER

Getting additional traffic means additional bandwidth consumption, which can cause issues. Flip to Chapter 10 for tips on ways to reduce the load on your servers and for info about optional hosting plans that don't charge for additional bandwidth.

Show Note Etiquette

Several schools of thought exist on how to approach show notes. Some podcasters say you should be very brief, using notes only to hold URLs and other pieces of important offline data that your listeners may not have had time to write down as the show was playing. Others suggest show notes should be filled with information on each and every concept touched upon in the show. Whether you prefer a more moderate approach or a deeper dive into the format of your show, your personal tastes and style go a long way in determining what is right for you.

Setting aside the level of detail you want to explain, you need to follow some basic rules of etiquette:

>> **Use intriguing and informative titles.** In general, and to keep things simple, the title of your show notes should match your episode title. Your title is your pitch; you're a huckster competing for the attention of listeners. Some listeners may know all about you; others could be seeing something from you for the first time. Include keywords in your title that accurately and specifically represent the contents of this episode. Your keywords should also generate some excitement and make the episode sound interesting and intriguing to potential listeners.

WARNING

There is a fine line between "intriguing and informative" and "clickbait" when it comes to titles and headlines. Don't go for the sensational and avoid headlines that could come across as accusatory or toxic.

>> **Include links to resources mentioned in the podcast.** If you're talking about a trip to the local museum, provide a link to the museum's main site in your show notes. If you mention another podcast, link to it. If you mention a news story or opinion piece, drop that URL into your show notes. Don't forget about music credits and affiliate or sponsor links too! Good linking brings good karma, and it may provide some interesting and potentially helpful "Hey, you linked to me!" comments (and backlinks) from others.

>> **Concise or complete?** Show notes can be as simple as a bullet list of topics or as detailed as a word-for-word transcript of your show. It's up to you. The advantage to the bullet list is that it's quicker to put together, obviously. If you're already scripting your podcast, it's no extra trouble to post the script, but be aware of the length. A transcript for a 3-minute show is pretty easy for someone to read from their browser; however a 45-minute transcript — yes, there are 45-minute shows that are completely scripted — may be something you don't want the reader to go through.

Figure 11-1 displays how Tee composes his show notes for *Happy Hour from the Tower* (http://happyhourfromthetower.com). Each episode includes backlinks, embedded videos, and bullet points about the topics covered. Is it a lot of work? It can be, but the end result is a guide that listeners can easily follow.

FIGURE 11-1:
The Fireteam
from *Happy Hour
from the Tower*
implement show
notes that range
from simple
backlinks to
embedded videos
related to the
conversations.

Planning the Post

The amount of time you spend planning your show notes is inversely proportional to the amount of time you spend during your show prep (see Chapters 5, 6, and 7). If you forgot everything from your high school math class, allow us to paraphrase: The more you prepare for your show, the less time you spend working on the show notes — and vice versa.

Examine the notes you used when you recorded your podcast. Did you talk about any websites? Find the URLs and make sure you spell them right. Test them. Make sure they are headed to the right place. We highly recommend the copy-and-paste technique for URLs, rather than relying on your typing skills, especially for lengthy URLs.

If you recorded and/or edited your show hours or days before you started this notation process, replaying the media file with pen and paper at the ready is a good idea. Look for need-to-know moments and jot them down as the show plays. After it finishes, use a search engine to find additional, relevant URLs you may want to provide to your listeners.

It's all in the details

Now is a good time to figure out what level of detail you're going to employ in your show notes. Several factors can influence your decision, and audience expectation and personal choice are among the more important.

TIP

Here's a good rule: The deeper you dive into a single topic on your podcast, the less detailed your notes need to be. That may sound counterintuitive, and please keep in mind this is only a general rule and not a law. For example, if your podcast episode features a 20-minute interview with Theoretical Physicist Dr. Michio Kaku on his book *Physics of the Impossible* and how applicable those ideas are to the *Star Wars* Universe, you likely won't have much more than a link to buy the book and/or rent the movie.

Show note details serve several primary purposes:

>> To act as a table of contents for the episode

>> To allow listeners to skip ahead if they so choose

As the podcaster, you can decide how much or how little you embrace these purposes. Here are some approaches that other podcasters have adopted:

>> **Add a time stamp on segment or topic changes.** Some podcasters put the exact time stamp of when they change topics, which can be frequent depending on the show's format and its host. Time stamps can be quite helpful to your listeners if you cover a wide range of topics in a given episode and want to assist possible listeners in jumping around. Keep in mind, though — time stamps will take longer to note. Yes, it's a terrific touch, but it's a commitment.

>> **Write in complete sentences and paragraphs.** Taking cues from the world of blogging, many podcasters, such as Michael J. Rigg does in *Steamrollers Adventure Podcast* (http://riggstories.com/the-podcast), write show notes in prose, using complete sentences and paragraphs in place of bullet points and time stamps. This approach feels better to potential readers, giving them a flavor of the show without having to listen. However, we've also heard listeners complain that key elements are difficult to find in this format. The prose approach works best for fiction-driven and RPG podcasts or for short, quick podcasts designed for business.

>> **Create a simple one-line summary.** Some podcasters, such as Manoush Zomorod and her NPR podcast *Note to Self* (https://www.wnyc.org/shows/notetoself), take a minimalist approach and post simple one-liners or maybe three sentences that quickly sum up what the show will cover. We

suggest new podcasters not follow this lead because it doesn't do much for helping attract new listeners. Many shows that take a quick summary approach enjoy a wider distribution method. In this case, *Note to Self* is also broadcast over WNYC93.9FM.

REMEMBER

Detailed show notes improve your search engine rankings, and they enable curious Google visitors to determine the value of an episode before listening. Show downloads and subscriptions can spike whenever your episode touches on hot topics related to your podcast as these keywords appear in show notes as well as blogpost tags. That's how powerful show notes can be.

A picture is worth a thousand words

Podcasters will also include a representative image or two in their show notes, as seen in Figure 11-1 and in 11-2. While random graphics serve only to increase your bandwidth consumption and risk cluttering your page, well-selected images can add flavor and dimension to your show notes.

Before you add an image to your post, keep in mind these three considerations:

>> **Is the image protected by copyright?** Simply grabbing an image off the Internet can land even the most well-meaning podcaster in a heap of legal trouble. Remember in Chapter 5, the section "I hear music. . ." about music in your podcast? Same thing. Here's a few options to help avoid copyright issues:

- *Plan A — Use your own images.* No issues there because you own them and can use them any way you like.

- *Plan B — Use Google Images* (https://images.google.com). When you search, go under Tools ⇨ Usage Rights and select Creative Commons license. These images are the ones companies and personalities make available for public use.

- *Plan C — Identify the owner and ask for permission.* Most people are eager to get their images displayed and are happy to say yes, provided they get credit.

>> **Can you link directly to the image, or do you need to copy it to your server?** Some sites, such as Amazon.com, allow you to link directly to images as they sit on the website. These sites have a huge technology infrastructure and can handle remote hosting images that appear on other sites. But many smaller and personal sites can't handle the load a popular podcast can put on their systems if they allowed direct linking to their stored images. In these

cases, copy the image to your own server before adding it to your page. If you're going to do this, it's good karma to provide an "image courtesy of . . ." link to the original site. Again, this assumes you've received the appropriate permissions to copy the file. When in doubt, don't.

TIP

If you are using images related to your show — a podcast about electric vehicles, for example — and you use images of the latest Tesla model, that is more than okay as you are using images directly related to your show. However, if you are using images of Elon Musk to promote your baseball podcast (dropping speech bubbles on Musk saying, "When I'm not creating ships bound for Mars, I'm waiting for the next *Full Count* podcast!"), that is crossing a line.

» **Does the image fit on your page?** Images too small or too large aren't doing your listeners any favors. Make sure the image you select is the right size. You can add width="x" height="x" declarations to your image tags to control the size or use your blog engine's editing presets to resize the image (better to go larger-to-smaller than the opposite direction), but keep in mind that this might distort the image. Previewing your post with resized images is a must. If your HTML is a little rusty, check out *Coding For Dummies,* by Nikhil Abraham (Wiley), for additional help.

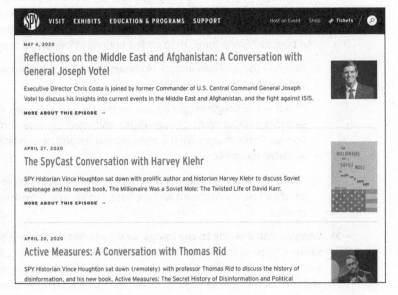

FIGURE 11-2:
For each episode of *The Spycast* (https://www.spymuseum.org/multimedia/spycast/), a different image is featured, giving episodes' show notes an extra distinction from one another.

Posting Your Show Notes

If you've planned and prepared, posting your show notes is easy. And if you've decided for the quick-and-dirty approach or don't really care to use show notes, this process can go quickly as well because there's nothing to do, right?

In this section, we show you how to enter your show notes by using WordPress and LibSyn as examples. If you use another tool to make your posts, or if you create your notes by hand, you still get value out of these examples as we show you things to consider along the way.

Posting in WordPress

WordPress is free, easy-to-use blogging software that also works well for podcasters. Follow these steps to post show notes in WordPress (https://www.wordpress.org):

1. **Log in to your website's site administration page.**

   By default, you typically can find a Login link on WordPress pages. If you've already provided credentials, the link may say Site Admin. In most instances, the URL looks something like http://www.*your_domain_here*.com/wp-admin.

2. **Choose Posts ➪ Add New from the menu along the left of your browser window, as shown in Figure 11-3, to start a new posting.**

   You can also choose +New from the menu across the top of your browser window. A new posting page opens. Here's where you fill in the details of your new posting. Although this screen displays a lot of items, you don't need to use them all to get started.

3. **Select the appropriate category for your podcast from the Categories section.**

   The Category feature keeps your various posts organized. It's not uncommon to have one category for text/blog entries and another for podcasts. For example, you can have a category specific for Podcast Episodes and another category dedicated to Latest News where it is a blog entry completely independent from your podcast.

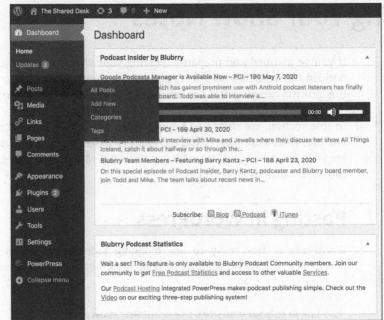

FIGURE 11-3:
Once you log into the WordPress interface, you can begin a new post either by going to the +New option from the top menu or the Posts option from the left-hand menu.

4. **Enter the title of your podcast in the Title text block.**

   WordPress uses *text blocks* to compose your blogposts. The text blocks offer a variety of options for you. Some of the text blocks' functions include:

   - Title
   - Text
   - List
   - YouTube embed
   - Twitter embed

   When you launch WordPress, every post begins with a Title block and a Text block. You build your post from here. For more details on the WordPress user interface and how it works, take a look at *WordPress For Dummies,* 8th Edition by Lisa Sabin-Wilson.

5. **In the text block offered, enter your show notes.**

   Follow a chronological order and list the various topics covered in your show, one on each line.

6. **Be sure to add URLs to any websites you mention.**

   To create a link, highlight the text you want to link and then click the Link button (the chain link icon) in the Text block's interface, pictured in Figure 11-4. Copy and paste the full URL — including the `http://` part — and then press your Enter/Return key.

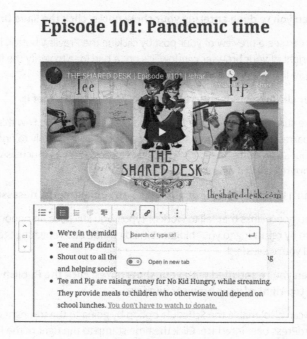

Episode 101: Pandemic time

7. **Connect your podcast file to the blog.**

   Before entering in your show notes, you will want to install a *plug-in*. Figure 11-5 shows Blubrry's PowerPress interface, but there are others out there you can install into WordPress. Your media file should be on your server (covered in Chapter 10), so use that pathname from your server and the plug-in will detect your file.

REMEMBER

What if you wanted to feature a podcast you were on? What is the best way to share that experience with your audience? Well, this is when you *syndicate* an episode. The episode appears in your feed, but the statistics go back to the original podcast creator. Podcast plug-ins make syndication incredibly easy. Instead of the media hosted on your server, it resides on another server, a *remote* server. Go to your plug-in and use the entire URL of the podcast where you are featured — for example, `http://theotherpodcast/episodes/path_to_yourinterview.mp3`.

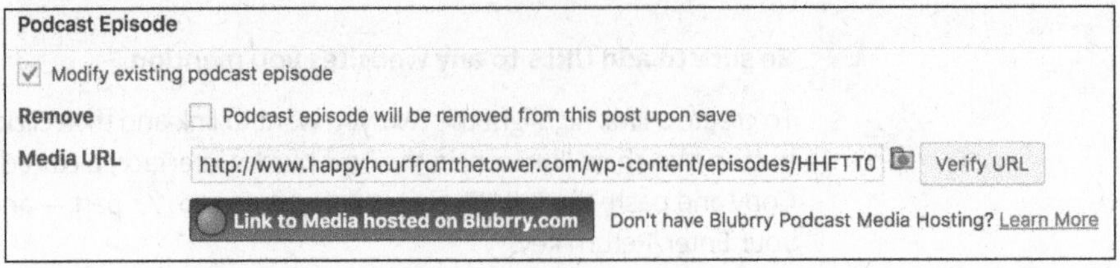

FIGURE 11-5:
Blubrry's
PowerPress
interface.

> **Podcast Episode**
>
> ☑ Modify existing podcast episode
>
> Remove ☐ Podcast episode will be removed from this post upon save
>
> Media URL http://www.happyhourfromthetower.com/wp-content/episodes/HHFTT0 📷 Verify URL
>
> 🔘 **Link to Media hosted on Blubrry.com** Don't have Blubrry Podcast Media Hosting? Learn More

8. When you're done entering your show notes, click the Save Draft button.

You can see a preview of your post by clicking the Preview button in the top-right of your browser window. It's not a bad idea to verify the format and ensure links will work before releasing it to the public.

9. After clicking Preview, scroll through your post to proof it.

Make sure links work properly, including your podcast file. Few things are more embarrassing than releasing that long-awaited podcast only to find that a link in your show notes doesn't work. It's up to you to test as much as you can before releasing a new episode.

10. Return to your post editor and make any adjustments necessary.

If you notice links that fail to work or typos, you still have an opportunity to fix them by returning to your blogpost in Edit mode and repeating Step 8 and 9 until you're satisfied.

11. When you're satisfied with your show notes, click the Publish button in the top-right of your browser window.

You can also enable the Schedule option by going to the top-right of the WordPress user interface. Click the time stamp to the right of the Publish option under Document ⇨ Status and Visibility and enter a date later in the week. You can use the Schedule option to cue shows for automatic posting.

12. Click View Site link at the top of the page to see how your entry looks.

Visiting your web page is a good idea to make sure everything looks as you expected it to. If it doesn't, simply edit the post and save your changes.

TIP

Many podcasters with detailed show notes use the <More> tag in their WordPress postings, the tool and tag highlighted in Figure 11-6. Any text above the More link is displayed on the main page. Those with interest to view the detailed content can click the link. If you have lots of show notes to include in your post, this is a good way to keep your front page tidy with a few bullet items for each show, and then you can "hide" the longer list.

FIGURE 11-6:
When creating show notes in WordPress, incorporating a *More* tag can help keep the appearance of your blog's main page tidy.

Posting on LibSyn

Because LibSyn (https://www.libsyn.com) is a dedicated podcast hosting provider, the steps are quite intuitive and a bit different than those used by folks who work with blogging software.

Follow these steps to post your show notes on LibSyn:

1. **To create a new post, under Content, select Add New Episode.**

   TIP

   LibSyn also allows text-only posts to the site, which is great for times when you want to post some text without a media file, such as to say "I'm on vacation for the next two weeks."

2. **Click the Add Media File button.**

3. **From the pop-up window, select Upload from Hard Drive and then select your episode file and click Choose.**

4. **Under the Details tab, enter a title for this episode.**

   We cover some titling tips later in this section. For now, simply enter a basic description of what this episode is about — "Classic Car Auction" will serve here as our example.

5. **Enter the detailed show notes in the Description box.**

   This is the appropriate place for your detailed show notes. Follow a chronological order and list the various topics covered in your show, one on each line. Be sure to add URLs to any websites you mention.

TIP

Typing the URL in this section isn't quite as effective as providing a true hyperlink. Making a hyperlink isn't difficult, though it may look that way if you aren't familiar with the mysteries of HTML.

Hyperlinks follow this convention:

```
<a href="link/to/website/or/web/page">Name of link</a>
```

Basically, you fill out what's between the quotes, replace the name of the link, and you're done. Here are a few real-world examples:

```
<a href="http://www.chuckchat.com/gmail">Gmail Podcast</a>
<a href="http://morevi.net">Morevi podcast</a>
<a href="http://marsrovers.nasa.gov/gallery/images.html" target="_
    blank">Pictures from Mars</a>
```

Take note of the target="_blank" part of the last link. By including that in the HTML tag, you are telling the link to open in a new browser tab.

6. **Choose a category for your podcast from the Category drop-down list.**

   Categories on LibSyn work like they do with most blogging applications. To keep this simple, choose Podcast from the list. This Category is only used with internal LibSyn tools, such as its web page, media player, and smartphone apps.

7. **On the Details tab, fill in the information for Apple Podcasts Optimization.**

   The key items to fill in are the Apple Podcasts Title, Episode Type (most cases it is Full), Episode Number, Rating, and Author.

8. **Under the Artwork tab, upload custom episode level artwork (if you have any).**

TIP

   Because LibSyn doesn't allow you to preview your post first, we highly recommend checking your spelling and double-checking your links before you proceed. An ounce of prevention and all that.

   Figure 11-7 shows how your post looks on LibSyn before it is posted.

9. **Under the Scheduling tab, pick the specific day and time you want the episode to go live or just select Publish.**

   LibSyn makes it hard to screw up, requiring you to fill out the appropriate fields before allowing you to continue. If you're successful, you see a pop-up screen where you can copy the direct link to your media file and/or the embed code to the player for that episode.

10. **Check your page.**

    Click the link to your show name and see how it looks in the real world. Check your links, spelling, and layout. If it's not the way you want, close the window and click the name of your post to edit it.

FIGURE 11-7:
Filling out show notes on LibSyn is an easy process with an intuitive UI.

TIP

It never hurts to mention in your show that you have show notes. You can do so as part of the running dialog. For example, in the middle of recording, say ". . . We found a great deal on those at Frobozco. We'll have a link in the show notes on our website . . ." or mention it at the end of the show with contact information, as in "Don't forget to stop by our website, where you'll find links to everything we mentioned in the show notes, information on how to contact us, and much more at www. . . ."

Boosting Search Engine Rankings with Good Show Notes

One tangible benefit of quality show notes is the impact they can have on your listings within search engines. Traditional search engines cannot (yet) scan and index the contents of your podcast media files. As such, you need to provide text for search engines to examine and evaluate for index inclusion.

TECHNICAL STUFF

Podcasters can pick up a lot of tips and tricks from bloggers and other website owners on how to boost search engine rankings. Many include page-level changes to positioning of elements, correct usage of headings, meta and image tags, and back-linking techniques. That conversation is far beyond the scope of this book, so grab a copy of *SEO For Dummies*, 7th Edition by Peter Kent (published by Wiley) if you want to make a bigger splash.

In the following sections, we show you some best practices you can implement right away that can make your notes (and podcast) more accessible to search engines — and ultimately search engine users.

Loading up your titles

Search engines (and searchers) pay close attention to titles. You should consider the title of your individual podcast episodes every bit as important as the title of a given web page.

Important as they are, most podcasters struggle with effective titles. The biggest problem comes from confusing titles with descriptions. If your title starts with "In this episode," stop right there. You're writing the description, not the title. A title is a string of well-chosen and crafted words that has no room for superfluous baggage.

We find that the best titles come from a re-examination of your show notes. If you haven't made your notes yet, you may find coming up with a solid title quite tough. Here's the process we suggest:

1. **Read your show notes and pull out the key elements, thoughts, or themes covered on the show.**

   Let's say you have a podcast on a Classic Car Auction. In your show notes, you have covered:

   - Sleeping in the Seattle airport while the flight was delayed

   - Interesting discussion during the flight with an 80-year-old man who is a car restorer

   - Under the hood of the 1965 Ford Mustang

   - Custom headers and exhaust systems

   - A short interview with the owner of a 1972 Chevy Nova

   - Taking a 1957 Chevy Belair for a cruise

   - Listener feedback request: What's your favorite classic car

2. **Boil down each element to a single word or phrase, if possible.**

Think about the people who might be interested in the contents of your show and pick common words they're likely to search for.

Potential episode titles from your bullet list above:

- Sleepless in the Seattle Airport
- The Old Man and the Chevy
- Two Decades of Drive
- Making a Classic Your Own

These all make solid titles, giving searchers a good tease about what they can expect. It all depends on what kind of title you want for your episode. A sense of humor? Alliteration? Something more literal? Whether your audience is more casual or, in the case of our Classic Car Auction podcast, more serious collectors and restorators, you select the key points from your show notes and condense them to what works.

REMEMBER

Notice how the title doesn't cover everything, and it shouldn't try. That's the job of the description where you can go into even greater detail on those three elements, plus the many other things you talk about on the show.

Titles also carry good keywords likely to be of interest to your audience. Chances are good that you know your audience much better than we do. Think about how people are likely to search and write your titles for that. Keep them short, don't try to cover everything, and employ more detailed descriptions to carry the rest of your story.

Soliciting backlinks

Backlinks are the Holy Grail of search engine optimization. A *backlink* is simply a link from someone else's website to yours as opposed to a link from your site to another. In this case, you are looking for outside locations to link back to your own site. Sites that have a lot of backlinks pointing to them are considered "more important" to the computers that control where your site shows up on a search engine. To get backlinks from others, you have to create links to their websites in your show notes.

TIP

When you're soliciting backlinks from sources, make each email personal, provide the exact link you want them to use, and tell them why you think it's important for them to link it.

Before you post your show, contact other podcasters, bloggers, and perhaps notable websites to see whether they're interested in linking to you (and you to them) to generate interest in your podcast, rather than a specific show. Many sites have a section somewhere for related sites. This can be a very effective tool for drawing people to your site and your podcast. Then you can also find folks who might want to backlink to particular episodes.

You've posted your show and got your show notes online. It's time to start soliciting backlinks to individual episodes:

>> **Company backlinks:** Write to the company that manufactures the custom headers to explain that you're posting a review of one of its products. Getting big companies to link to you doesn't always happen, but sometimes it does. And getting backlinks from big popular sites is very beneficial to your rankings.

>> **Courtesy backlinks:** If someone helped you with a part of your podcast and you mention it in your episode, let that source know. Send an email to the agency you booked the trip with and maybe even the hotel.

>> **Backlinks from fellow podcasters:** If other podcasters cover topics related to your episode, let them know about it because they might be willing to spare you a backlink. Notify various car bloggers and podcasters about the new episode. It takes only a few moments of your time and is information they may welcome.

WARNING

A fine line exists between asking for backlinks and spamming someone. If you can't think of a good reason why that site should link to you, then you don't ask for it; otherwise, just mention what your show is and what you've covered that might be of interest and let them decide whether they want to provide a backlink.

4

Start Spreadin' the News about Your Podcast

Chapter **12**

Speaking Directly to Your Peeps

ommunication can be defined in a multitude of overly complex ways. For the sake of argument (and not to copy each and every dictionary entry we can find), we define the term this way:

The exchange of information between two points.

Note that last part — between two points. To us, this implies a bidirectional flow of information, to and from both parties.

If you've had the pleasure (note how well we can say that with a straight face) of attending any productivity or team-building seminars, the presenters really drive the message home: Effective communication is not a one-way street.

Over the past few years, podcasting has evolved into a more effective communication method than traditional media (such as radio or television). We all have communication tools at our disposal — email, websites, phone lines — so why do podcast listeners seem to get more involved with podcasting? Two simple reasons:

» There seems to be a closer bond between podcast consumer and podcast producer. The simple fact that anyone can do this makes the producer seem more like a real person than a personality and therefore, easier to relate to.

>> The podcasters are asking for feedback — and getting it. Audience size doesn't matter. We've seen some instances of shows with a couple hundred loyal followers where the podcaster has to spin off a second show just to handle the listener feedback.

In this chapter, we show you some real-world examples of how to foster communication between you and your audience, touching on a variety of methods and venues.

Gathering Listener Feedback

It must be a natural human reaction to fear the opinions of others. Perhaps it's insecurity, but we think it has more to do with our culture's constant reinforcement of the "How are you?" — "I'm fine. You?" — "Fine." meaningless chatter that precedes most of our conversations.

That cultural crutch, however, is left next to Tiny Tim's seat when it comes to podcasting. Listeners, for whatever reason, are compelled to actually give real and meaningful criticism. And podcasters, for the most part, take to heart those responses.

Of course, we're speaking in general terms. Yes, there are flamers and trolls out there with less than helpful opinions at the ready. Podcasting can't change basic human nature for the ill-evolved, unfortunately.

You can foster good communication with your listening audience in a multitude of ways, such as:

>> Allowing and responding to comments on your blog

>> Creating and visiting online discussion groups and forums

>> Responding to listener email

>> Participating in online social networks

>> Leveraging voicemail, as seen in Figure 12-1

So let's get cracking on how we make these methods of reaching out and touching someone work for your podcast.

FIGURE 12-1:
When a show goes live, remind your audience on social media that you have a voicemail line and that you'd love to hear from them.

Fostering Comments on Your Blog

In an ideal world, all communication, feedback, rants, and raves about your show would take place in a neat little box, keeping things nice and tidy for you. But because that won't happen, your best bet is to build a website that both enables and encourages the communication in your own backyard.

In Chapter 11, we demonstrate how much adding show notes can improve reaching new audiences on account of SEO. Well, here's one more reason show notes are important to your podcast — show notes serve as mechanisms for interaction.

In the world of blogs, this interaction is referred to as *comments*. Visit just about any blog you can find, read a post, and you're likely to find a small Comments link at the bottom. Some podcasters get dozens of responses per episode. Some get none. Although there is some relation to the size of your listening audience, the frequency of your podcast, and the number of comments you're likely to receive, it really has more to do with the connection users feel they have with your podcast.

REMEMBER

If you're already using a blog, you usually don't have to do anything special to turn on the comments feature. Most software comes configured to accept comments by default. If you decide comments are not for you or your podcast, turning off that capability is simply a matter of selecting the right option.

Much like two co-workers chatting about last night's *The Late Show with Stephen Colbert*, two or more listeners using your website to talk to each other about your show speaks to the attachment they feel to what you have to say (or play). From the very first moments of attachment, you can nurture a community around your podcast. Here are some simple ways we recommend to foster your following:

>> Mention that you have comments on the blog as a standard practice as part of the intro, outro, or the break.

>> Include a one-line promotion of your podcast in your email signature.

>> Ask questions during the show and direct listeners to your blog to leave their comments or opinions.

>> Actively respond to received comments.

Communication develops amongst the listening community itself. Rather than talking to you, listeners start talking to each other, and the conversation — and sometimes, the community — takes on a life of its own.

WARNING

Sometimes, comments that take on lives of their own mutate into hostile take-overs, and the conversation is lost amidst an onslaught of personal attacks either to other posters or to you, the podcast's and blog's host. Neither your community nor you need this kind of conversation. You may want to consider *moderating* blog comments. Moderating still allows people to post comments, but you get to approve those comments for public consumption. When someone posts a comment, you get an email. With a couple mouse clicks, you can chose whether to approve it. You have to do a little more work, but moderating ensures all comments meet your quality standards.

You may find your name or podcast mentioned other places on the Internet via search engines (discussed both in this chapter and Chapter 11). We recommend you get engaged with those discussions as well. Be sure to check back in a couple days for additional follow-ups, or if you are responding to a blog comment or forum, check to see if they have an option to alert you to replies.

Focusing on Online Forums

An *online forum* allows individuals to post their thoughts and ideas on a variety of topics — of their own choosing. Through a concept known as *threading*, multiple discussions can exist independently of all the others. Topics can get buried quickly in email discussions. Forums work differently, keeping all threads and topics available for clutter-free commenting at any time.

REMEMBER

As easy as it is to create places for these types of conversations to occur, someone else may have already done it. Spend some quality time searching the Internet for your name and your show. Maybe a devoted fan has already done the not-so-heavy lifting for you.

Using a hosted forum, like a hosted blog, takes away much of the burden of downloading, installing, and configuring the software. On the other hand, you don't have quite the flexibility with a hosted solution as you would with a package you host yourself. One of the easier hosted forums to use is Tapatalk, as seen in Figure 12-2. To get started, just follow these steps:

1. **Browse to www.tapatalk.com.**

If you have a Tapatalk account, enter your username and password.

If you don't have an account, click the Sign Up button.

2. **Log in.**

Click either Google or Facebook to log in with one of those accounts, or click Continue With Email to create a new account using your email. If you choose the last option, check your email to validate the email address.

3. **Create a new group.**

Click the Start New link on the Start Your Own section. Enter a group name (your podcast name). The short name automatically fills in. Provide a short description of your show, pick a color, and optionally upload a background image. Validate that you are not a robot and click Continue.

FIGURE 12-2:
Tapatalk offers free forum hosting that you can use for your podcasts.

4. **Configure your group.**

On the home page under Groups, locate your group, click the three vertical dots (the kebab icon), and modify the group settings to your needs.

5. **Send invitations.**

Using the same etiquette as a distribution list, let the world know your forums are available. You can send an email or announce it on your podcast.

Keep in mind the issue of how some people might see invitations as spam.

That's it! You can now customize your forum, start new posts, and spread the word of your newly created forum. Much like discussion groups, you can promote your forum by

>> Posting the address on your website

>> Adding the address to your email signature line

>> Mentioning the forum on each of your podcast episodes

You can find plenty of hosted forums options by running a Google search. Most are supported by web ads or voluntary donations. Moderating a forum takes three to four hours a week for troubleshooting, answering user requests, and keeping the spammers and forum troublemakers at bay. Four hours is a modest investment considering the return is a strong community.

However, keep in mind that forums and communities take time to build steam. Be persistent, post every day, and constantly encourage your listeners to interact with you and your podcast in this manner. But above all, be patient!

Remember when we said podcasting has an element of marketing with it? Here's another opportunity to take your name to the people. Developing a public forum means your information becomes available to a lot of potential listeners. They may find your forum and become interested in your podcast, rather than the other way around.

Like blog comments, you might find your name or podcast being thrown around other forums. We highly suggest you sign up and participate in the discussion. Respond to the comments — positive or negative. (See the section about not-so-positive comments later in this chapter.)

Social Media

Social media has become a part of our lives over the past decade and a half. Online communities built around these platforms are made up of profiles, posts, tweets, pages, instas, and snaps for people of all ages and backgrounds. They can serve as one-stop shops, or combine popular forms of communication into a hybrid that gives you all the best features. Some of these platforms have been online longer than podcasting has been around, but podcasters rely on these platforms as a ways and means of reaching new audiences.

We cover many of the more popular networks out there. Yes, there are more than the ones highlighted here, but these are the ones getting a lot of the attention.

REMEMBER

You can find plenty of social networking sites out there, and many more will probably appear (and disappear) in the time it takes this book to reach your hands. What's important is finding the network or networks that can get the word of your listeners back to you, help you create a better show, and make you a better podcaster.

Facebook

The house that Zuckerberg built, Facebook (www.facebook.com) has evolved from the "alternative to MySpace" (yeah, remember when MySpace was cool?) to a one-stop shop offering a blog, bulletin board, online scrapbook, video streaming, and forum, all in one convenient location. Facebook gives registered users the ability to create online *Groups* where comments can be circulated across others' networks as well as your own. You can also establish a *Page* where you control the message being sent out to the public and boost posts that will appear in Facebook News Feeds around the world. Facebook provides your listeners a quick interface both through your computer and your mobile device to not only offer feedback on what just went live, but also preview what's coming up next on your podcast. You can also use your Groups and Pages to solicit voicemail or offer up polls that serve as instant content for your show.

TIP

Facebook also offers Facebook Live (as we talk about in Chapter 17), a video streaming service where your mobile device's camera captures where you are and what you are thinking. While you are filming, comments and reactions are shared with you and your audience. If you are sharing video in your podcast feed, you can easily download your video and drop it into your feed. If you would prefer not to offer video, then extract the audio from your Facebook Live segment and use it as a new podcast episode. Facebook Live can be used as either a promotional device for your podcast, or as content for upcoming episodes.

For more information on Facebook, pick up a copy of *Facebook For Dummies* by Carolyn Abram.

Twitter

Twitter (www.twitter.com) continues to be a fantastic and instantaneous way of beginning conversations, garnering feedback, and getting word out about your podcast while not becoming a distraction or productivity time-sink as Facebook sometimes tends to be. Twitter, either through its website, its mobile app, or a third-party application that is Twitter-enabled, gives you 280 characters to say anything. You can use Twitter to direct people to your blog when a new show posts. Listeners can post (or *tweet*) what they're listening to and comment on it. From various tweets, topics can be created on the blog or forums, resources can be cited, and quick announcements can reach a wide variety of listeners in moments.

Twitter delivers the instant gratification of tweeting but gives you only 280 characters to do it, preventing you from losing your intent in a drawn-out posting. It's based on the premise of answering the question "What's happening?" If you find someone is listening to your podcast, ask them for feedback. Good or bad, begin a simple chat and ask for the opinions of others in your Twitter network. Use Twitter to post teasers on upcoming episodes, ask for validation from comments found elsewhere, and tweet relevant links either you or your listeners provide that tie back to your most recent episode's topic.

TIP

Before Facebook Live, Twitter developed its own video streaming platform — Periscope (www.periscope.tv) — allowing for viewers all over the world the opportunity to comment on what you are sharing. As we discuss in Chapter 17, your video in Periscope can be saved to your phone, and then uploaded either to your feed or your video platform of choice. If you would prefer not to offer video, then extract the audio from your Periscope segment, and drop it as a new podcast episode. Periscope can be used as either a promotional device for your podcast, or to provide content for upcoming episodes.

Pinterest

Known more as a haven for D.I.Y. projects or slow cooker recipes, Pinterest (www.pinterest.com) is a platform waiting for you and your podcasting street team to take full advantage of. When you understand how Pinterest works, podcasters can tap into its full potential. Think of the social network as a visual bulletin board and every time you post a new show, you go to your virtual bulletin board and post an image relevant to your show. That could be your show art, or it could be an image from your show notes. Your fans go to this board, click the image you just posted, and they find themselves on your podcast's site. That's how *boards* on Pinterest work.

Set a podcast board for your show, and then, when putting together show notes, go on and incorporate relevant images for whatever you are talking about. When you create a new post — Pinterest calls this a *pin* — Pinterest will ask where to pull images from. Use your latest episode's URL and then pick an image to represent the new content. Others in your Pinterest network can now interact easily with your pin by either leaving a comment or repinning it to their boards, reaching a whole new network.

YouTube? For audio?!

"My podcast is audio-only. What possible use could I have for YouTube?" Yeah, that's what many podcasters think — they couldn't be more wrong. The challenge is that YouTube (www.youtube.com) doesn't allow uploading of MP3 files. Everything has to be a video. It's not a big leap to use something like iMovie, Premiere Elements, or Screenflow to take your audio track, slap in an image and create a video file you can put on YouTube. You've already done the editing for the audio file (if you're in to that sort of thing). Now it's just a couple of additional steps to import and export to upload to YouTube (see Figure 12-3). Don't short yourself: Take a look at Chapter 17 for some useful bits about creating a simple video file.

FIGURE 12-3: Ben and Keith of TGGeeks distribute every episode on YouTube.

What's the point of putting your audio-only podcast on YouTube? Simple, you want to be where the people are. Billions of people are on YouTube all looking for content. Suppose your podcast is on Japanese history. . . People are searching for information on Japanese history on YouTube. Sure, it may not be your biggest distribution channel, but like the other sections in this chapter, it's all about making yourself as visible as possible.

Instagram

Instagram (www.instagram.com) may not come to mind as a promotional platform or communication channel for your podcast, but with some ingenious approaches, the image-exclusive platform gives your podcast an exciting new way to let people know that new episodes are live and how to send feedback through voicemail.

So how do you turn an app all about capturing the moment visually into a community platform for your *audio* podcast? It may require a few workarounds, but once you find your rhythm, it becomes second nature after a few postings.

Once your latest episode goes live:

1. **Mail your Show Art to your smartphone.**

   At the time of this writing, there are no apps that allow you to upload photos from your computer to your Instagram account. Instagram was always meant for smartphones.

2. **Save the Show Art into your smartphone's Photo app.**

3. **Pull up the URL of your show in your smartphone's browser, find the new episode's URL, and copy it to your phone's clipboard.**

4. **Launch Instagram and go to your Instagram profile by tapping your profile icon in the lower-right corner of the app's Options.**

5. **Tap the Edit Profile option.**

6. **Paste the episode's URL in to the Website field on your Instagram profile (highlighted in Figure 12-4).**

   You can use the main URL for the podcast, but the individual episode's URL will take your audience directly to the new episode.

7. **Tap Done to accept and activate the changes.**

8. **Tap the Create Post option (the + icon) in the Instagram menu.**

9. **Create a new Instagram post with your Show Art as the featured image; make sure to include in the post "Follow the URL in my Instagram profile . . ." so people know where to find the new content.**

   URLs are not active in Instagram posts.

REMEMBER

WARNING

If you are creating an Instagram profile just for your podcast, you do not want an Instagram account that is nothing but images of your Show Art. You will want to either create Instagram Show Art that feature images relevant to your podcast's content or post other content that may be in tune to your interests, or even the show's interest. Reporting the same image over and over again could get you reported as SPAM and, in turn, shut down.

Now that you have your podcast featured on Instagram, how about giving your listeners an easy way to leave you voicemail? They are already on their phones. How can you make that happen?

1. **Launch Instagram and go to your Instagram profile by tapping your profile icon in the lower-right corner of the app's Options.**

2. **Tap the Edit Profile button to access the Account Options menu.**

3. **Scroll down to the Switch to Professional Account option. Tap this option and follow the steps to identify as a Creator account.**

   At the time of this writing, there were several categories to identify you. While Podcaster is not yet one of the categories, we recommend you choose *Digital Creator* as a close second. When you complete this and go back to your profile screen, you should see some new settings at the bottom.

4. **Tap the Page option to connect your profile to a new or existing Facebook page then click Done.**

5. **Tap the Contact options to set an email address and phone number so your audience can easily reach you, then tap the checkmark when done.**

6. **When you've completed your profile settings, click the checkmark on the Edit Profile screen to return to the profile screen.**

   On finishing this process, notice that your Instagram profile now comes with a Contact button. By tapping this button, visitors to your Instagram profile can directly reach out to you, as shown in Figure 12-5.

FIGURE 12-5: With a simple tap, listeners of your podcast can now leave you either email or voicemail feedback regarding your latest show.

Social media offers you these options and a whole lot more. These platforms are a breeze to set up, but you know what else is easy-peasy to set up? Voicemail. If you have a smartphone and if you can navigate through Google, you are only a few clicks away from your own voicemail account.

Discord: The Lava Lamp of Online Communications

Yes, we cover online forums and social media as great ways to cultivate your community, but back in our third edition, there was a new player that we didn't know about. This new player in communication, as you see in the this section's heading, is something like a lava lamp: stylish, always changing, and synonymous with bringing people together for a variety of reasons.

Discord (`www.discordapp.com`) is available as a smartphone app, a tablet app, and a stand-alone desktop client, and hosts millions of daily users and hundreds of millions of messages a day. While Discord is more associated with console gaming, eSports, and streamers (something we touch on in Chapter 17), podcasters are turning to the platform as a solution in building a community. It is part online forum, part social media, and something entirely unique.

By establishing a *server* on Discord, you offer one location for your listeners or viewers to meet. Once on your server, you and your audience can keep in touch through *text channels* and *voice channels,* designated by hashtags as seen in Figure 12-6. Discord offers you a unique ability to not only develop and build a community around your podcast, but also keep the conversation rolling in real-time, offering fans of the show a special online *kaffeeklatsch* (a German term that translates to "coffee and gossip" or friends getting together at someone's house for a good old fashioned chin wag) in a special meeting room.

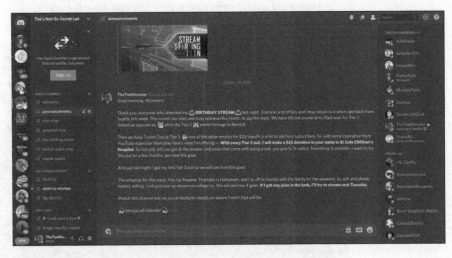

FIGURE 12-6:
Discord is a versatile, reliable communications platform, and is considered the premier community builder for content creators.

And with the right apps and consent from those attending, you can record the Discord gatherings and offer them up as podcasts themselves. Think of these shows as live feedback roundtables.

In this exercise, we are going from the perspective that you already have a Discord account, server, and channel set up; and we're going to make a post about our latest podcast episode. If you need to know how to set up a Discord account, set up a server, and then establish a channel, take a look at Tee's *Discord For Dummies* title. He'll get you up and running.

If you're already running, you can widen that stride of yours:

1. **Go to your #general channel and select it.**

2. **Enter in the following text into the message field:**

   A new episode of _The Shared Desk_ is up and **LIVE** right now, so have a listen at `http://www.theshareddesk.com/2020/06/02/episode-102/`.

   We welcome creative couple Alyson Grauer and Drew Mierzejewski who are podcasting _Skyjacks' Courier's Call_ but this isn't their first rodeo. Have a listen and share your thoughts at **703.791.1701**!

   TIP To get a hard return/line break in a Discord message, use Shift + Enter on a desktop keyboard.

3. **Once you have your message typed out, hit the Enter key or the Send icon to post your message.**

   This quick message, shown in Figure 12-7, lets members of your server know that your latest episode is now live and ready for their ears. Sure, if listeners are subscribed to your podcast, they will be notified by their podcast app of choice but a little reminder in your Discord doesn't hurt.

FIGURE 12-7: Along with continuing discussions about recent episodes, Discord can also serve as a notification system to help you get the word out that a new show is live!

What you should also notice in your recent post are the different kinds of text formatting available. Discord offers you the following:

» _italics_: Placing text between a pair of underscores will italicize text. You can also italicize text by using a single asterisk on either side of your text.

» **bold**: When you use double asterisks, any text appearing between the two pair will be bolded.

» ***bold and italics***: For additional emphasis, you can simultaneously bold and italicize text by using a set of three asterisks on either side of the text you wish to format.

» ~strikethrough~: Strikethrough can always be fun when you want to show a sudden change of thought or illustrate how a change in one draft can differ from another. To do this, you surround the stricken text with a *tilde*, created by using the Shift key and the key to the left of your "1" key.

» _underlining_: Similar to italics, you can underline text by using a pair of underscores on either side of a body of text.

WARNING

Old habits die hard. When people see underlined text, it is perceived that whatever is underlined is a link. Even though links are not underlined in Discord, it is something that happens often. Underlined text might earn you a few postings of "Hey, do you know this link is broken?" from visitors. Also, underlined text can sometimes be hard to read. Use the underline markup sparingly, and at your own risk.

Then there are *emojis* available just by clicking on the smiley face off to the right. Drop a laughing face, and people know you are kidding. Drop an angry face, and your intent is made clear. Yeah, this may come across as a weird detail to point out, but there is a good reason why you have so many emojis to choose from. It's a final touch along with basic text formatting to help you fine-tune your tone, so consider using them.

Discord offers you a terrific opportunity to build a community around your podcast. Take a look at it, either through your own server or in participating on other Discord servers, to see whether this platform works for you in connecting with listeners.

Using Voicemail

One of the strengths of podcasting is that the content is so portable. That means your listener is quite likely to be away from a computer while listening to your podcast — making the interaction more difficult. Plenty of people, including your authors, listen to podcasts during their commute to work or on road trips. How do you get comments from those listeners? Simple: Have them call in.

Mobile phones are practically ubiquitous. It's a fair assumption that your listener has a mobile phone and sometimes they are listening to your podcast on their mobile phone. If you want those listeners to give you feedback, give them a number to call. Many podcasters have set up a number through Google Voice (https://voice.google.com). Google Voice, as seen in Figure 12-8, lets people call and leave a message.

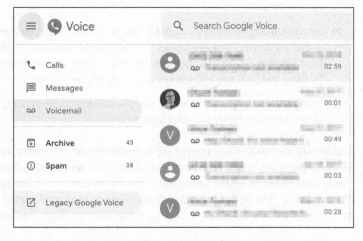

To set up a listener line, follow these steps:

1. **Sign up for a free Gmail account** (https://mail.google.com) **if you do not already have one.**

2. **Go to** https://voice.google.com.

3. **Click the gear icon in the upper right.**

4. **Click Account on the left menu.**

5. **Click the Choose button in the Choose a Google Voice number section.**

6. **Review the Terms of Service and Privacy Policy and then click Continue.**

7. **Enter the city or area code for which you would like to get a free phone number.**

   Note: This number does not have to be in your area.

TIP

Pick a phone number that is easy to remember or has a mnemonic, a memory pattern, you can choose, for example 1234, or perhaps a number that correspond to letters. If you are doing a *Star Trek* podcast, a number with 8735 (TREK) somewhere in it makes it easy for your callers to remember.

8. **Click Next to link your Google voice number with your mobile number.**

    Make certain you enable the Do Not Disturb option for your Google Voice number. This will automatically take any calls going to your Google Voice number straight to voicemail.

9. **Google will send you a verification code. Enter this on the screen presented and click Verify.**

10. **When the process is completed, click Finish.**

You can now tell the world about an easy way to leave you a voicemail. When someone leaves you a message, you'll get an email notification!

Okay, Google Voice is fine and dandy if you're calling from the United States, but let's not forget that podcasting can reach anyone anywhere. How do you get in touch with the listeners from Australia, South Africa, or Tierra del Fuego? You can encourage your international audience to leave you a voice memo through Facebook Messenger or take advantage of WhatsApp. Their audio can come from anywhere in the world for free.

You may find yourself with so much voicemail that you'll have to do what several podcasters have done and create a separate show for their listener feedback. Listeners, including your authors, often get a kick out of hearing their comments aired to the world.

Seeking Out the Comments of Others

There's an old saying about the best-laid plans of mice and men (and how they often go awry). That adage can be applied quite aptly to when podcasters pick up a walking stick, throw a haversack over their shoulders, and proclaim, "I'm going on an adventure!" You see, it is a certainty that listeners of your show, both fan and foe, will talk about your show to others in a variety of formats and on platforms of which you have absolutely no control.

There are existing forums, chat rooms, and social media threads that deal with your particular podcasting topic. At some point, those people will find out about your show and start listening. Current research shows these people will post reviews faster than Ken Miles's now legendary 1966 Le Mans race. In fact, uneducated opinions on the Internet stand as the *only* things faster than Miles's legendary Ford GT40.

Welcome to the community of the Internet.

There are a variety of ways to keep your eyes and ears on these groups and to find comments regarding your podcast. Doing so will give you valuable, direct feedback from listeners and let you respond quickly and easily. But before you set off on that journey, consider the warnings passed on to Indiana Jones before setting off on a archeological quest: Be careful what you unearth.

Trying a general search

Is it just us, or don't most people do a Google search for their own name at least twice a week? Could be just us, but that's a great way to see whether people are talking about you. Google has a gazillion pages in its search database and constantly crawls a good percentage of the web, finding interesting tidbits and adding more data with each pass.

When you search, try various combinations. If your name is a common one, such as *John Smith*, you're probably going to get a lot of hits unrelated to you. Try adding the topic of your show to the search for more relevant results. For example, if your name is John Smith and you're podcasting about underwater basket weaving, type John Smith underwater basket weaving in the Google search box. If your show name is unique, or at least uncommon, try using the name of your show as a search term.

TIP

We realize that there are other search engines besides Google. Yahoo! and Bing produce fine results, as do a few others. If you have neither the time nor the inclination to experiment with a dozen search engines, we suggest these three. They syndicate their results to other lesser-known (but equally valid) search engines. But as we've said countless times before, your mileage may vary. The same techniques we outline work well on just about any search engine you prefer.

Searching within a site, blog, or social media platform

As extensive and cool as search engines are, they can't cover everything on the Net. Not only are there physical limitations as to how wide of an area the spiders and bots can cover, there are also self-imposed limitations set up by website owners that inhibit a good indexing of the site. Take forums, for example. Some are set up in a manner that renders their internal pages invisible to the spiders and bots of even the best engines.

But most forums have an internal search engine that you can use to find the content within the forum — though you may be required to register with the forum to access its search engine. Blogs, forums, and even social media platforms such as Twitter have search features; the results are easy to track down, as shown in Figure 12-9.

When the Comments Are Less than Good

First, don't panic.

Second, don't respond. Not yet.

Third, let your blood pressure come down to a normal level.

Let's face it. Anytime someone has any critical comments about us, we get an emotional reaction. We call that being human, and it's perfectly understandable and impossible to suppress. Following that impulse of replying right away only leads you to discover the two reasons why it's called a *knee-jerk reaction*: It's a reflex to clashing viewpoints, and you come across like a real jerk when you don't think about your response before riding the emotional roller coaster. (The classic wooden ones like the *Rebel Yell*, *Beast*, or *Grizzly*. Yeah. Roller coasters. Cool.)

When you're calm and feeling a bit more detached, reread the comment and plan your course of action. Here are some suggestions:

1. **Reflect on the comment.**

   What does the comment say, really? Does the person make a valid point? Is there an area of improvement you should make? If the comment was specific, re-listen to the show in question. Did you say what the person said you did or stumble as bad as the person made it out to be? You may need a different perspective, so feel free to get someone else involved.

2. **When you fully understand the criticism, decide whether or not you want to respond.**

   One option is to not respond at all. If you do choose to respond, consider taking the high road as the best approach. You can often disarm an inflammatory remark simply by politely accepting the comments with something like "Thank you for your feedback. I'll take it in to consideration."

3. **Additionally, consider sending an email.**

   If you send an email, count on that email being posted right alongside the negative comment. There's no guarantee the person will keep your correspondence private. In fact, count on the opposite. Whatever you say in a private email should be something you would be willing to say in a more public forum. Keep your rebuttal rational, civil, and, above all, professional.

REMEMBER

Just like any argument, it's best to keep things on a professional level with your words and mannerisms. Although it may be hard not to take it personally, try to refrain from escalating. It's not going to help the situation.

Negative feedback is never an easy thing to stomach, but look at the positive aspect of this: People are listening. They're listening, and now they're most assuredly talking, blogging, and podcasting about you. We're not saying to rush out and say something completely irrational simply to drum up controversy, but we're saying that people will disagree with you now and then. It should be expected, and you should be ready to face that tough love when it comes your way.

More than anything, grow from the experience. Understand that anything you say in your podcast will be heard by a variety of people, with different backgrounds, experiences, and expectations of the world. We've been on both sides of this and can count many times where the negative comments we received turned out to be some of the best feedback. We think we're better podcasters for it.

Feedback, good or bad, is only as constructive as you make it out to be.

AVOID THE BAITING GAME

Tee recalls, after watching a scene in *Bull Durham* where an umpire goads Kevin Costner into insulting him, saying, "That's so Hollywood." His dad, a highly respected umpire in college and semi-pro baseball, said, "Actually, that's not. That's *baiting*. I've played that game too many times." Although you may not think there is a common trait between baseball umpires and podcasters, there is: The Baiting Game.

Fast-forward to 2005 when on Tee's podcast *The Survival Guide to Writing Fantasy*, he posted a tribute to a friend lost to breast cancer. Billed as a "Special Edition" this show was simply to raise awareness. An anonymous poster came on to the show's blog and left a long comment that said in so many words, "We don't care about your friend dying of cancer. You're a professional writer so be professional. Stick to the content of the show." While Tee did not reply, his community of "Survivalists" did, to which the anonymous poster rebutted with a lot of contempt and very little reason. On his following *Survival Guide*, Tee asked the poster to stop listening if he found the show so distasteful. The anonymous poster returned to say "You can't make me stop listening . . ." and then listed other writing podcasts he found better than Tee's.

At this point, Tee recognized this for The Baiting Game that it was.

There are people out there who live to pick fights. They are commonly referred to as *Trolls*. Trolls can be "listeners" of your podcasts or, sometimes, other podcasters looking for a thrill in bucking the community. A way to tell it's The Baiting Game is to look closer at the criticism with (pardon the pun) a critical eye. Usually the feedback from baiting is comprised of personal attacks, misconstruing of facts, flawed reasoning (if there's any reason present), and a healthy dose of verbal insults akin to punching someone in the ear and then running away. When this happens, don't reply. Ever. By not replying, the words sent to you are a waste of time and effort on the sender's part, not yours. Not replying isn't a guarantee that the baiter will simply disappear. You may need to moderate your blog just to be sure people don't post comments that you deem inappropriate.

And as you may be moderating said comments, you can continue to delete the trolls' comments, and they never see the light of day.

The high road has a better view than the low. In some cases, like The Baiting Game, the best response is none at all.

Chapter **13**

Fishing for Listeners

A s of this writing, there are hundreds of thousands of podcasters. By some estimates, the number of subscriptions exceeds one billion. However, that pales in comparison with the potential audience, which is somewhere in the billions of people worldwide with a broadband Internet connection. Although the audience is large, the options for listeners are legion. But how do you attract an audience to your podcast?

In this chapter, we show you a variety of options that you may want to use to help gain a larger listener base. Some cost money; others cost time. But using these options, you can expose your podcast to the right people at the right time.

Getting Your Podcast Ready for Promotion

Whether you plan on spending real money or expending real energy, you need to do some prep work before you start your campaign. You shouldn't rush promotional campaigns — take the time to carefully plan and execute them. Failure to do so can not only be a huge waste of time and money, but it may also result in turning off potential listeners to your show, making it many times more difficult to attract them back for a second chance.

Polishing your presentation

Most podcasters need a few shows under their belts before they hit their stride. If you're on podcast episode number three, you likely haven't fleshed out your show. Granted, you may have been planning your podcast for months on end, or have previous experience behind the mic on another medium, or have nailed it from the beginning. If so, great. But understand you're in the minority. And much like a new TV series with experienced writers, producers, and directors, experienced podcasters launching a new show should consider releasing a few shows until that show is running like a machine.

REMEMBER

Even though each person is different, we suggest giving yourself at least five full shows to find the sweet spot. Experimentation is part of the format, so play with a few things along the way to see where your strengths as well as your weaknesses are.

Checking your bandwidth

In Chapter 10, we introduce you to the concepts of unmetered bandwidth and unlimited space. Each new listener means more of your precious bandwidth being consumed. For podcasters with limited bandwidth, getting more listeners can be an expensive proposition.

REMEMBER

If you're using the services of Liberated Syndication (http://libsyn.com) or another unmetered bandwidth podcast hosting company, you don't have to worry about your bandwidth and can safely skip ahead to the next section. See Chapter 10 for more information about these very affordable services.

Many podcasters start out using the standard web hosting service to host their podcast files and are quickly surprised when they run out of bandwidth. We're more surprised about how poor their math skills are. Suppose that you have 100MB of monthly transfers allowed for your site. On the 10th of July, you log in to your bandwidth stats page and see that you're already at 60MB for the month. Will you make it? Here are the formulas to figure this out:

Bandwidth consumed / number of days so far this month = Daily bandwidth rate

Daily bandwidth rate × 31 (the total number of days in the month) = Total bandwidth needed

Now you plug in your numbers to find out your daily bandwidth rate:

60MB / 10 = 6MB

You're consuming about 6MB per day. Now multiply this number (6MB) by the number of days in the month (31) to get the total bandwidth you need for the month:

6MB × 31 = 186MB

It's inevitable; you're not going to stay within your 100MB limit. You will be roughly 86MB over your 100MB plan.

That sounds — and is — simple. As a real-world example, Chuck checked his bandwidth usage on July 18 and found that he had used 256.65GB in the first 17 days.

First, he needs to find out his daily bandwidth:

256.65GB / 17 days = 15GB average daily transfer rate for July

Then, he has to find out his total bandwidth for the month:

15GB × 31 days = 465GB

As you can see, he needs 465GB of bandwidth to get through the month, assuming his traffic stays steady and doesn't increase. If his bandwidth ceiling was 500GB, he'd need to think twice before starting an advertising campaign, as he'd likely hit that ceiling, and his hosting provider would likely shut down access to all those brand-new podcast listeners he just worked so hard to get. Not a good way for anyone to spend his or her time.

As a good rule, you need to be using less than 50 percent of your monthly allotment of bandwidth before starting an advertising campaign. If you're using any more than that, you'll run out of room and will have to seek alternative hosting options before proceeding. We cover some of these alternative options in Chapter 10.

Figuring out your USP

USP is a marketing term, and it stands for *unique selling proposition* — a message that sells your podcast to potentials listeners. Although you probably aren't charging money to listen to your podcast, make no mistake that you need to "sell it" to potential listeners if you're considering advertising.

Why should a potential listener listen to your podcast? And more importantly, how can you, as the podcast advertiser, present a message that makes a potential listener want to listen?

Plenty of books, websites, seminars, and post-graduate degree programs are dedicated to the subtle nuances of marketing and advertising. We're not suggesting you go that far, obviously. But we do suggest you take a good, hard look at what you produce every week and come up with a concise and consistent message with which to promote your show.

For example, when Chuck was looking at advertising for his *Technorama* podcast, he found that the market for another tech podcast was pretty saturated. Several dozen shows repurposed the same news about Microsoft, Google, and Apple. Instead, Chuck chose to spotlight the strange, bizarre, and unusual items that are typically passed via email from geek to geek — the steam-powered Nintendo DS, the motorcycle that folds in to a briefcase, and who can forget the device made of wood that adds binary numbers. Thus, the USP for *Technorama* goes like this:

> *Technorama* takes a light-hearted look at the world of tech, science, sci-fi, and all things geek.

TIP

Sometimes calling in help from the outside can be a good thing. Ask your friends and family, or even your listeners, to come up with some key points of why they listen to your show. We're not talking about a catchy slogan or jingle of the sort a Madison Avenue marketing firm might designate as the "perfect" thing to attract new listeners, but plain English (or your language of choice) ways to tell interested folks what your show is about and why they should be listening.

We can't give you a step-by-step outline on this one. Spend a few days on it. Try it out on some folks first. When you find a message that fits, you're ready to proceed on your advertising quest.

Exploring Various Advertising Options

Tee and Chuck have been podcasting for over 15 years, and advertising is best described as somewhat daunting. There is nothing "easy" about this because there are no surefire solutions. It's similar to gambling. The odds could very well be in your favor today, but tomorrow is a completely different set of rules. There are many ways you can advertise to give your podcast exposure to a larger audience. Yeah, it can be a little intimidating. That doesn't mean you can't take a chance or two, and see what happens, right?

IS ADVERTISING RIGHT FOR YOUR PODCAST?

Before launching your advertising campaign, apply some *Jurassic Park* logic. In that movie, Jeff Goldblum's character chastises the dinosaur-resurrecting mogul with "Your scientists were so preoccupied with whether or not they could, they didn't stop to think if they should." Good advice for podcasters and mad scientists alike.

For many, podcasting is a labor of love and not a money-making proposition. As such, spending too much money or time on advertising may make a passion seem a heck of a lot more like a job — and you probably already have one of those.

For years, Dave Slusher's *Evil Genius Chronicles* (www.evilgeniuschronicles.org) was very vocal about not advertising and not working toward a huge listener base. Dave makes a podcast for one person — Dave Slusher. If other folks hear about the podcast and decide to listen, great, so long as they enjoy it and don't expect him to be something he's not. Fame is a double-edged sword — even the moderate fame a popular podcaster can achieve. Before long, emails and voicemails are flying in with ideas, suggestions, and even mandates of how you can make the show better — for the listeners. Don't forget that *fan* is short for *fanatic,* and fanatic people don't always behave rationally.

Choose this path with caution. "Doing it for the love" probably means you can (and should) forgo advertising your podcast. It's not a fast way to get more listeners, but it is a fast way to burn through money without a lot to show for it.

If you're interested in making money on your podcast, the first thing you will want to consider is the *Return on Investment,* or *ROI.* Now we're starting to sound more like *Advertising For Dummies,* so bear with us a moment. Simply put, is the money you spend on advertising going to generate you any additional. . . anything? It doesn't make sense to spend $5 or $5,000 if you won't get more. . . something — listeners, advertisers, sponsors, money — in return. This is ROI in a nutshell: making more money than what you invest because spending money just to spend money doesn't sound like a lot of fun to us.

Give me a boost, Facebook

Still the largest of social media platforms, Facebook is a good place to consider a modest financial investment in order to reach a larger group people with hopes of converting them into loyal listeners. This is the same reason you see ads for shaving razors, car batteries, or pillows to help you stop snoring on your Facebook feed. Someone wants you to see what they have to offer and somewhere in your massive collection of data, you fit the profile — so why not turn the tables and start reaching people who fit your profile?

How it works

Facebook boosts allow you to expose, for a price, your Facebook post on the timelines of people who don't directly follow you or those who may have shared your content. The price is determined by how many people you want to reach.

For the sake of argument, let's assume you've already set up a Facebook page for your podcast as one of the many ways you can stay in touch with your people and your people can stay in touch with you. As you post a new episode, you are mentioning it on your podcast's Facebook page, right? But let's say your Facebook page has 30 followers. It's great that those 30 people know about your new episode, but you want to reach more, so you share it via your personal account to your 500 followers (mostly family, friends, and co-workers) in hopes that some of them share it. How do you reach more people you *haven't* already met to let them know? The equivalent of going to a party and shaking hands and introducing yourself and your show? That's where Facebook helps. They allow you to "boost" a post for a few dollars and define whom you wish to target.

TIP

Boost a post containing your episode, not your page. People will be more interested in hearing your breakthrough research on stopping telomere breakdown to extend a cell's lifespan than just knowing about your science research show.

Using the service

You should have no problem finding your post to boost. Facebook practically pushes it in your face when you view your own timeline. Chuck found it odd at first to see *Technorama* ads with this latest post on his timeline. This is Facebook's subtle way of increasing the likelihood of you paying them. Very clever! Then Chuck looked closely at the top of the ad that says "Only you can see this preview until you run this ad . . ." and a blue button at the bottom that says Boost Post.

TIP

Save your *boost budget* for your special shows. Like most podcasters, you are working on a budget and want to save it for when it really matters. If you do a music podcast and just landed a really spectacular interview with Alice Cooper, that's a good candidate to boost!

When you click the Boost Post button a window pops up, as shown in Figure 13-1.

FIGURE 13-1:
The Facebook
Boost Post
feature lets you
target Facebook
users with
interests that
align with your
show's content.

Facebook gives you a few ways to choose your target audience for your ad:

>> When you click the option for **People you choose through targeting**, you'll also notice an Edit button that allows you to pick specific genders, age range, locations, and topics of interest. Think of the topics of interest as your way to match keyword searches with people. If you just did an episode with a noted independent horror film director, you might include things like Independent movies, horror movies, movies, and other related items that might capture the interest of Facebook users.

>> The option **People who like your page** doesn't sound like it's going to get you anything more than people who already like your page and are likely to see your content, but thanks to Facebook's magical algorithms, it also targets people who like other pages similar to yours.

>> **People who like your page and their friends** lets you place your ad not only on the timeline of people who already follow your page, but their friends as well. It's almost like paying everyone to share your link for you — does that make you feel dirty too?

>> And the final option: The **Create New** link is similar to the first option, but you can create groupings of targets to save and use on multiple ads. This is handy if your show has a variety of topics on a regular basis. Let's say you do a science podcast: You might set up one grouping that targets people interested in space exploration, NASA, and Mars, while another ad would be better

targeted to DNA, gene sequencing, and genetic mutations. That way when you finish that interview with the rocket technician, you can target the right group and repurpose that group weeks or months later when you air the tour at JPL.

Of course, there's always the subject of money. How many people can you reach and how long do you run the ad? Fortunately, you can choose a budget that works for you, starting with just a few dollars and running for as little as a day. The more you spend, the larger potential audience you can reach. The longer you run the ad, the more it will cost you. The pop-up window has a special section that helps you calculate the cost before you commit. No surprises here that will force you in to a second mortgage, thank goodness.

Finally, there's the method of payment. Facebook makes this pretty easy by taking credit or debit cards, PayPal, and online banking accounts. Pretty straightforward.

WARNING

Be sure to read the Facebook Advertising Policy (https://www.facebook.com/policies/ads) before launching your campaign. Yes, it's a lot of text, but it can save you from nasty surprises that get your ad rejected. Chuck learned this on his first attempt when he discovered that images on boosted posts must not contain more than 20% text — frustrating.

While Facebook is a major player in social media, this may change over time. We encourage you to look into other avenues for ad campaigns like PayPerPost (https://payperpost.com/).

Insta-traffic with Instagram

In the Chapter 12, we walk you through the process of switching your Instagram profile from a personal account to a business profile. With a business profile, you can offer your listeners and viewership quick and easy access to your podcast's email and voicemail.

Now we're going to show you exactly how you can do business on your Instagram business profile.

In 2016, Instagram — with a lot of help from its parent company, Facebook — offered to business account holders the option to boost Instagram posts. You can easily take one of your images and turn it into a sponsored post, reaching beyond your own network to Instagrammers that may not be following you, who share interests relevant to your podcast.

When you activate your (free) business profile, Instagram connects to either a Facebook account or a Facebook Page that you manage. Along with the option to go to a profile and tap a Contact button, all your posts will offer you the option to promote that post. When you *promote* a post, your Instagram post — the image and its accompanying text — appears in feeds of audiences you target. If this sounds a lot like what happens when boosting Facebook posts, it's because you are working with the same audiences you are building with Facebook.

And promoting a post on Instagram is just as easy as on Facebook:

1. **Launch your Instagram app.**

   Your Instagram business account must be activated to complete these steps. See Chapter 12 for the steps for how to do so.

2. **Create a post pertaining to your podcast or find a post about your podcast in your Instagram profile.**

3. **Tap the blue Promote button.**

4. **From the Select Where to Send People options offered, tap** Your website **to set this promotion for a specific episode.**

   This is called an *Action Item* — what you want your audience to do when they see your ad. Most of the time, you will want your audience to visit a URL.

5. **Tap the arrow in the upper right to continue.**

6. **Select your Audience, either from audiences you have built in Facebook, walking through steps to build your own audience, or having Instagram do it for you.**

   REMEMBER

   When you sync your Instagram business account with a Facebook Page, any audiences you have defined on Facebook are automatically carried over to Instagram.

7. **Tap the arrow in the upper right again to set your Total Budget and Duration for this ad campaign and then proceed to the next screen.**

8. **Tap the Shop Now option associated with the Action Button option; select Learn More as your button.**

   You can now review your order at this screen and enter a payment method. The ad campaign can be previewed here (as shown in Figure 13-2) or submitted for review by tapping the big blue Create Promotion button.

After you send your Instagram post, your post is reviewed by Instagram (just as Facebook does when you boost a post there), and you are notified when a promotion goes live. Instagram will then keep running analytics on your post, keeping you informed regarding how much traffic you are seeing, what your *CTR (click-through rate)* is, and how much remains in your budget.

FIGURE 13-2:
Before you submit an Instagram post for review, you are given a chance to review your order before making it live.

TIP

When you set up a boosted Facebook post, you are given an option to also simultaneously run a boosted post on Instagram, provided the post you are boosting is an image with an appropriate resolution. Make sure you use an image that looks good on a laptop as well as a mobile device.

Instagram's analytics on a boosted post, you will notice, are kept separate from the analytics you collect on the original post. In other words, you will see that some people liking your boosted Instagram post do not add numbers to the original post. This is Instagram's way of keeping organic traffic separate from paid traffic. Once again, it falls on you to decide, based on your budget, how many people you want to reach, how long to run the ad, and if you need to change the demographic. Presently, Instagram's budget suggestions for posts are modest, if not humble. Choose a budget that works for you, starting with just a few dollars a day and having it run for a week. As with Facebook, the larger your budget and the longer you run it, the larger your *potential* audience.

TIP

Regardless if you are using the "Promote" option or simply using Instagram to promote your podcast, consider the accompanying post and your Profile's URL. As we mention earlier, you cannot make active links in Instagram (yet). Before posting your show's art or image related to your episode, go into your Instagram profile and paste the URL of your latest episode into your featured URL field. Once you have the episode's URL in place (see Figure 13-3), go on and compose a post with "Visit the URL in our Instagram's profile . . ." somewhere in there. This way, people can reach your podcast through Instagram easily.

FIGURE 13-3:
Dropping the URL into your Instagram profile offers your audience a way to reach your podcast through the app.

REMEMBER

The hard and fast rule Facebook adheres to, that images on boosted posts must not contain more than 20 percent text, does not apply to Instagram posts. This is subject to change but presently show art like the one in Figure 13-1 is more than okay to use, provided it is at a good resolution like 600 x 600, for example.

Writing press releases

Some podcasters have enjoyed enormous success attracting new listeners with *press releases*. When you release a press release, it's available to be picked up by a variety of news sources. On the positive side, you have a lot of room inside a press release to talk about your show. On the negative, you have no guarantee of whom — if anyone — will pick up and run your press release.

Writing an effective press release is a true art form. It has to appeal both to the managing editor of the publication considering your release, as well as being worthwhile reading to John Q. Public.

Here are some ideas to help you write a good press release:

» **Hook them early.** This is also known as B.L.U.F. — stating your bottom line up front. Whatever you have to say, be sure to say it first and grab their attention.

» **Tell a story.** Your press release should convey a condensed version of what it is you want to say. Three or four paragraphs made up of two or three sentences each is a good guideline to follow. Try to answer the most common questions people will have about your announcement. Start with the basics of good journalism — who, what, when, where, and why.

» **Target the press release.** Write the press release suitable to your target audience. You will likely have to write more than one version — and send your story to those likely to run it. A story about "hometown boy writes successful book on podcasting" is more likely to get published in the college alumni school paper than *USA Today*.

» **Give the tag line.** Use your USP. This is the perfect place to give someone that 30-second pitch about your show.

» **Give some quotes.** Include a couple lines from the people involved. Keep them positive and upbeat.

» **Provide appropriate URLs.** Don't forget to provide links to your site(s). This should go without saying, but sometimes you may get so wrapped up in the wording of the press release that you forget the point of the press release — to get people to come to your site and try your content.

TIP

We recommend reading a number of press releases to get a feeling for the marketing aspect.

If you know people who have written press releases in the past, see if they can lend some help — if not, reach out. Remember that podcasting can be a very helpful network.

PR Newswire

PR Newswire (https://prnewswire.com) is the cream of the crop when it comes to online and offline distribution of your press release. PR Newswire, seen in Figure 13-4, has the ability to send your press release to thousands of media outlets, ranging from newsprint, radio, websites, television, and more.

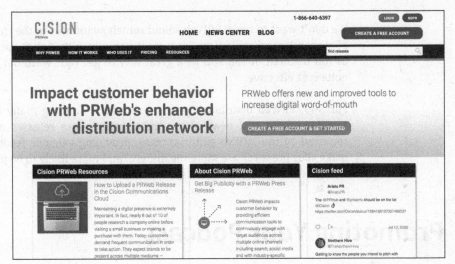

FIGURE 13-4: If you are looking to get the word out about an upcoming high-profile episode for your podcast, *PR Newswire* is there to help.

The company also provides editorial services for your press release, as well as tips and tricks on how to write an effective release that's more likely to get results. If you're serious about getting the maximum exposure to your show and are thinking about using a press release as part of your strategy, PR Newswire is worth considering.

However, many independent podcasters may find the cost prohibitive (around $300 to start).

PRWeb

Considerably less archaic is PRWeb (http://prweb.com). This organization offers paid and donation-based services, and it boasts of a good number of media outlets. The main difference between the two organizations (other than cost and the hassle-factor) is the means of the distribution. PRWeb works to put your press release in front of thousands of online media outlets. Offline media sources may subscribe as well, but the primary distribution is online.

Not that that is a bad thing. After all, your podcast is an online service, and you're likely most effective reaching an audience who is already online. Both your authors have had various write-ups in online and offline media, and we've always received more traffic and attention to our websites from the online sources.

We don't want to make PRWeb sound somehow inferior to the traditional sources of press release distribution. Far from it, actually. Although the service may not be as full featured, it can still be a great way to get your word out, and one that is quite cost effective.

For $99, PRWeb distributes your press release (among many other things) in search engines and news sites. Additionally, press releases at this level are reviewed by the PRWeb editorial team, providing valuable feedback on ways to make your press release more meaningful and more likely to be picked up by various other news organizations around the Internet.

Promoting Your Podcast

Before you start throwing money toward an advertising campaign, consider all the things you can do to spread the word about your podcast that don't require a financial investment. In fact, you probably should be doing these things even if you plan on tossing out some cash for effect.

Optimizing your site for search engines

In Chapter 11, we give you several tips on how you can make show notes more appealing to search engines. You can employ similar methods to your entire website so that it receives the maximum exposure and visibility to search engines.

For more information on optimizing your site, we highly recommend *SEO For Dummies* by Peter Kent (Wiley) and *Search Engine Optimization All-in-One For Dummies* by Bruce Clay (also by Wiley).

Submitting promos to other podcasts

Podcasting has been called by some to be a great hall of mirrors, as it seems that many podcasts spend at least some portion of their time talking about . . . other podcasts!

We think this feeling of community adds to the distinctiveness that is podcasting, and something that is, for the most part, accepted by the general podcasting audience. The podcasting landscape isn't shrinking anytime soon, and it's so fractured that many listeners are looking for their favorite podcasters to help steer them toward other podcasts they may find interesting.

Not all podcasters do this. In fact, the majority of the corporate podcasts must see other podcasts as competition and are as likely to talk about another podcast as a traditional broadcaster is to talk about another station across town. Even some of the independent podcasters make a point not to talk about other podcasts, simply because they don't want to add to the hall of mirrors effect.

One of the more widespread ways podcasters talk about other podcasters is with promos. A *promo* is an audio clip that describes your show. Other podcasters then insert this clip into their shows — play it on the air so to speak — thereby presenting your message to their subscribed and downloading audiences.

Promos are a great way to let other folks know your podcast exists. Spend some time listening to other podcasts and see whether they're playing promos. When you find one that does, make note of the average length of promos they play, and what type of content is being presented. Is it all serious business, or is more light-hearted humor involved? Query other podcasters and offer up your promo. In exchange, offer to play a promo for those podcasters gracious enough to play your promo. Some podcasters are really good at making promos and even offer to make promos for free for other podcasters.

One final note on sending out promos: Ask. Unless the show specifically says "Submit your promos to us at . . .," be a good podcasting citizen and send the podcaster a note asking whether she'd like to run your promo. Requests that start off with "I listen to your show every week because you . . ." are likely to get a better response than those starting (and ending) with "Please run my promo." Consider taking a reverse approach, as well. If you listen to a podcast and want to run their promo on your show, ask if they have a promo and offer to play it. Most podcasters are more than happy to contribute.

Recording your promo

Recording your own promo isn't difficult, and we tend to enjoy the ones that come from the voice of the podcaster. You've done the hard work by figuring out what makes your podcast special; now you need to sit down and record your promo. Here are a few tips:

>> **Write your script.** Or don't. Some folks are happy flying off the cuff. But in the interest of time, we highly recommend putting some thoughts down on paper and running through them out loud to see how long it takes. Most promos are under a minute long, unless you have lots of great stuff to say.

» **Add effects and music from your podcast.** If you use the same music (see Chapter 8) in your show each week or have some special sound effects that brand the show as yours, include them in your promo. Effects are a great way to tie in your promotion to your show, giving new listeners assurance they have subscribed to the right place.

TIP

Keep your promo between 30 to 90 seconds in running time. Anything over 90 seconds may be a hard sell for other podcasts to run it.

» **Don't forget your website URL!** Too many podcasters provide the link to their podcast feed. Useless, in our opinions. Instead, repeat the URL of your website, where it should be painfully simple to subscribe to the RSS feed for your podcast. Now here's to hoping you picked an easy-to-remember domain name!

» **Include a link to the promo on your website.** Recording one and sending it out to a few podcasts is great, but what about all the other folks who you inspire to make their own podcast? Chances are good that if you put a link to your promo file or create a page on your website, as seen in Figure 13-5, others will grab that file and include your promo in their shows. It's also a good repository for when you find another podcast you think might want to run your promo.

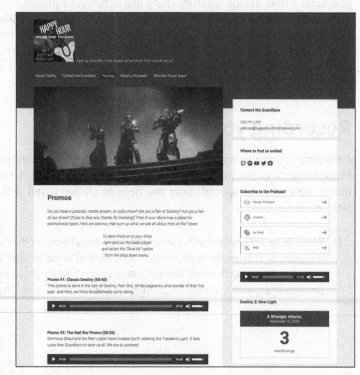

FIGURE 13-5:
If you choose to create promos for your show, make them easy to find on your site, easy to sample and download, with different run times.

>> **Speak clearly and enunciate!** You may know the name of your show very well. Like your own name, you've said it "millions" of times. This means you might say it so quickly, it may not come out as clearly as it should. When you introduce your show to someone who has never heard the show name before, you need to slow down a bit. Say your podcast's title as if you are introducing yourself to someone new. Let them hear the words clearly. While doing research for this book, we listened to several promos where the podcaster said the name of the show so fast that it reminded us of those mis-heard lyrics in a song. Was that Jimmy Hendrix podcast actually called "Kiss this guy dot com"? If it weren't for a printed URL, we wouldn't have any idea what their show was actually called or where to find it.

TIP

When you send other podcasters your promo, save some email bandwidth and send the link that's on your website, rather than the file itself. Make it easy to download the promo file from your site!

Giving interviews

Don't forget that podcasting is the fastest growing medium we've seen . . . ever! If your podcast covers a brand-new area of the world or addresses an underserved market, folks are out there who want to talk with you about it.

Contact the publications, radio shows, websites, and other outlets that cover the industry your podcast falls under. Send them your press release, along with a personalized note telling them about your show and stating that you're happy to do an interview.

Interviews can be done in person, but most today are conducted over the phone or online phone services like Skype. A handful are conducted via email. Preparing for an interview can make the difference between a poor interview that never sees publication or airtime and a well-delivered interview that keeps the audience — as well as the interviewer — engaged and entertained.

Here are a few tips to make your interview go swimmingly:

>> **Eliminate the BS factor.** If you have only a passing interest in the subject for which you're trying to pass yourself off as an expert, you'll quickly be discovered, thrashed repeatedly, and left out for the buzzards. The people who are interviewing you likely are already experts in their fields, so don't try to come off as something you aren't. Be open and honest about your experience and focus on why you're doing the podcast. Even if you're considered a subject matter expert, remember that your listener may not be. You might want to consider simplifying things on behalf of the listeners new to the subject. Of course, this depends on your listeners and the subject being covered.

>> **Mention your website over and over again.** Remember that the listeners, readers, or viewers you are being interviewed in front of have no idea who you are and what you are about. This is your chance to sell yourself and your podcast. If your interviewer is good, he'll give you ample opportunity to mention your website and podcast. If not, it's up to you. Look for chances to drop the name and URL if necessary.

>> **Stay positive.** If the interviewer knows anything about podcasting, he'll likely ask questions on the future of podcasting, the death of radio, amateur versus professional, and all sorts of other controversial topics. Unless your podcast is about podcasting, we recommend rising above the din. This conversation doesn't serve you or the listening/reading/viewing audience well. Point out how your podcast addresses the issue and resist the temptation to get into an argument. Unless your podcast is about arguing with interviewers. In that case, go right ahead.

Generating buzz

All the processes we outline in this chapter are geared toward one thing: generating buzz for your podcast. The more folks you can get talking about your show, the better off you are. We don't buy the "any publicity is good publicity" line, but we do think that "Hey have you heard about . . .?" conversations among real people are the best form of advertising you can get.

Sometimes, you need to take the message to the masses. Find a discussion group or an online forum germane to your podcast's area of interest and start posting.

When you post, follow standard forum/community etiquette — don't start out with "Hey, I'm new to the group and have this great podcast!" Instead, listen in on the conversation, comment on a few threads, and get folks used to your voice before you hit them with the come-listen-to-my-podcast pitch. In fact, the best way to pitch your podcast is to never utter those words at all. Instead, offer up things like "Last week on my podcast, I covered the very thing you were talking about, Jill." It shows you're paying attention to the conversation and not just looking to spam a newsgroup or mailing list with your podcast URL.

WARNING

A very fine line exists between tasteful self-promotion and outright spamming. Generating buzz isn't the same as advertising. Advertising has its place, but most forums don't welcome it. If you can't decide whether your post contains too much advertising or not, it probably does. Discretion is the better part of valor in this instance.

5 Pod-sibilities to Consider for Your Show

Get the basics on how to attract listeners and generate revenue around your podcast.

Learn how to use your podcast as a promotional vehicle to promote an agenda, an art form, or an artist.

Launch a podcast because you believe passionately in something.

Introduce streaming media into your podcasting workflow.

Chapter **14**

Show Me the Money

Throughout this book, we show you novel and interesting ways to toss significant amount of coinage into the proverbial black hole of podcasting. Throwing money at your podcast can easily become a habit — and it likely should come with a warning from the Surgeon General or perhaps your accountant.

Hosting fees, bandwidth overages, shiny new microphones, music royalties and licensing fees, phone charges, travel expenses . . . the hard and soft costs of this little hobby of yours just might add up quickly.

That's why this chapter shows you some ways to offset a portion (or all) of these costs and perhaps even add a few dollars to your pocket while you explore your newfound passion. In case you anticipate a large following, we cover some ideas you can use to make podcasting a paying gig.

REMEMBER

Being a podcaster is a lot like being an actor, a writer, or a professional athlete. A few select individuals may make it big, but most folks just get by — and that can be okay. If you're reading this chapter first in hopes of getting a crash course in how to "get rich quick" with your podcast, you're about to be disappointed. If not, we can talk.

How Much Money Can You Make?

Most podcasters fall into one of three categories in terms of their audience size: small, medium, and large (well, yeah). Because of the very low barrier-to-entry — in effect, anyone can create a podcast — the smaller variety of podcaster will likely make up the bulk of the community for the foreseeable future.

Here's a closer look at the moneymaking opportunities for these three podcast categories:

>> **Small:** Roughly, a small podcast has under 1,000 listeners. Having a small audience size doesn't exclude you from drawing a revenue stream from your podcast. It likely limits the size of your potential revenues, but it doesn't mean you can't bring in at least some income.

REMEMBER

Small is a relative term, and we're not about to start tossing out audience-size statistics to draw a clear demarcation between small and medium. Small is also not a derogatory term; many podcasters enjoy the idea of keeping their community intimate. There's a certain comfort in the small podcasts, a charm that some would say is diminished as the size of the podcast's audience increases. Small can actually be an asset — some podcasts are so niche that they draw a small, but extremely loyal following. To a potential advertiser, you have a target audience.

>> **Medium:** When you have more than a handful of dedicated listeners (in the four-digit bracket with over 1,000 listeners), you find yourself in the medium category. This category affords you additional opportunities. For instance, corporations and advertisers may be more willing to consider placing ads or providing sponsorships.

However, you also find yourself in a more competitive marketplace, as other podcasters start fishing for monies to help offset their costs. Stepping upward in the ranks also means stepping up your game, and you may find yourself in an unfamiliar place — trying to develop a media kit that boosts your podcast above the din raised by all the other podcasts chasing the very same advertisers.

TIP

Creating an effective media kit, especially for a newly discovered marketplace such as podcasting, is actually a pretty important task. For more about the why, what, and how-to, see "Developing a media kit," later in this chapter.

>> **Large:** Breaking the barrier at the far end of the size spectrum is an elite set of podcasters who occupy the *large* category and are in the five-digit bracket with over 10,000 listeners. When it comes to making money with your podcast, size really does matter. With a huge audience base, you'll likely find

advertisers a lot easier to approach. You can present listener numbers that are more like what the advertisers are used to seeing in their more traditional media buys.

Outside the mainstream media (like NPR, CNN, and ESPN), there are the juggernauts that launch when the stars are aligned, and the dice continuously land naturally on 20. Shows like *Welcome to Night Vale, Serial,* and *Lore* spawned shows of similar formats, similar approaches. And then there is Joe Rogan of *The Joe Rogan Experience* (http://podcasts.joerogan.net) who scored a $100 million deal with Spotify for exclusive rights. While all this sounds tantalizing, it's no picnic for large podcasters. Along with competing against other podcasts inspired by them (which is a nice way to describe knock-offs), large podcasts still target a very narrow audience on typically one topic (or show) and probably won't rake in as much as a nationally syndicated radio program. Additionally, the numbers game can sometimes become an obsession, and content quality may suffer on account of that.

However, if you see that paradigm between podcasts and radio change, be ready to make that jump. Podcasts enjoy a unique flexibility over broadcast media and arguably a closer relationship with the audience. Additionally, advertisers grasp the appeal of podcasting and are now making overtures, as seen in Figure 14-1.

FIGURE 14-1: Large podcasts like the popular true crime podcast, *Serial,* are beginning to catch the attention of sponsors.

DOES YOUR PODCAST NEED TO PAY FOR ITSELF?

"You spent how much on your podcast?" You'll soon be hearing that or a question just like it with a different gizmo attached to the question mark. Your non-podcasting friends, colleagues, and perhaps even family will gaze in wonder at your apparent lack of good judgment as you seemingly spend money as if it grew on trees — trees, we say! — adding just one more piece to your previously professed "perfect" podcasting setup.

How can you, in good conscience, justify this outlay of cash without the financial backing of someone else? Surely if no one is willing to pay you to do this podcasting thing, you shouldn't be doing it at all, right?

In a word, wrong.

Let's take a hypothetical, yet not atypical, household scenario — the family that likes camping. They've got their initial investment of tents, stove, sleeping bags, perhaps RV or pop-up trailer, maybe an ATV or two. For those of us who prefer podcasting, even a high-end, mic, mixer, mic boom, pop-filter, and some cables are a lot cheaper! Then there's the ongoing expenses. The camping family has campsite fees, vehicle maintenance — those ATVs don't fix themselves — food, fuel, park permits, and much more. The podcaster has hosting fees which we know are a lot easier on the budget than vehicle maintenance. Don't even get us started about how much space it takes to store a canoe compared to an iPad! Take some time and, no doubt, you could probably come up with several other hobbies and passions to offer to this little exercise.

Now back to the camping-loving family . . . ask them "Have you ever thought about getting sponsorship to help pay for your love of the great outdoors?" The look on their faces would be priceless. (Have a camera ready.) Just because you love to podcast doesn't mean you *need* a sponsor.

Think of the amount of money you spend eating out each week. Or for movie tickets, theater presentations, concerts, and/or bar tabs. These are things you love and enjoy, not things you're trying to find others to pay for. Most podcasters will spend no more than a hundred bucks a month on their show — average over time — and that's stretching it. And for such a small investment that has such a high personal payoff . . . do you *need* to be paid off in cold hard cash to do it?

Convincing Advertisers to Give You Money

If the idea of begging for money sounds rather repulsive, good. If you have to resort to begging, you shouldn't be asking at all. People will only part with their cash if you give them a compelling reason. "Because I'm a poor podcaster" is not a compelling reason.

Whether your goal is to gain sponsors or sell advertising spots, you have to answer a very important question: What's in it for them? Why should someone else make a financial contribution to your show? There are a multitude of good reasons that you'll need to uncover, understand, and be able to explain if you hope to be successful in asking for funds to support your podcast.

Contrary to popular belief, there aren't nearly enough corporations so bursting-at-the-seams with unused advertising dollars that they'll welcome you with open arms when you approach them and say, "Hey! Wanna advertise on my podcast?"

Most corporations have an *advertising budget* (as in, a limit on what they can spend in promoting products or services). In nearly all circumstances, this budget is significantly smaller than the range of potential places they might be advertising, so they're looking for the most bang for their buck. This means you're competing with media — radio, television, and online platforms — with well-established pitches and presentations at the ready, regularly scheduled exposition events like PAX East/West, CES, TwitchCon (featured in Figure 14-2) and DEFCON, and print media like *Wired, Forbes, The New York Times,* and *The Washington Post.* All these entities are working to get that sought-after advertising dollar.

Before you start cold-calling possible sponsors, do a little homework. Spend some time, energy, and money developing something your potential advertisers can touch and see: a media kit and a rate sheet, which we describe in the next sections.

REMEMBER

Sometimes all you can do is put your best foot forward and make your pitch. Talk to the advertiser about your dedicated audience, and how loyal it is to your show. Breaking down the costs into per-listener numbers might help as well. Whatever your approach, don't try to compare your numbers with those of traditional outlets the advertiser is already working with. Instead, talk about how adding podcast advertising to the mix can enhance its current efforts, allowing the company to reach an audience that has turned away from traditional media sources in favor of this new medium.

Developing a media kit

A *media kit* is, in effect, a collection of marketing tools designed to awaken potential advertisers to their crying need to shell out big bucks to support your product or service (in this case, your podcast). The size of your media kit depends on many factors, among which are how much you want to spend and how much important stuff you think you have to say. It's not size you're striving for here; it's a compelling argument that you can present to your potential advertisers.

Media kits are most certainly not one size fits all. Your media kit should be representative of the actual feeling you try to produce on your podcast — competent and real, not over-the-top glitzy. Consider it like a job interview: It's okay to comb your hair and put on a fresh shirt, but you wouldn't send your good-looking-but-ignorant roommate as a stand-in, would you? Know how to clearly communicate what your podcast is all about and how it would benefit a potential sponsor.

REMEMBER

The big fish are very well armed in this pond. Large media outlets tend to have large media kits that were developed with the assistance of large advertising agencies that charged large sums of money for their expertise. The small- to midsize podcaster need not go this far. But you need something more than just a collection of local news articles about you and an online portfolio of your latest season.

So here's a practical list of dos and don'ts to keep in mind when considering what to put in your media kit.

First off, the "Yes, go for it" list:

>> **Include accurate listener statistics.** Okay . . . as accurate as you can get. Focus more on total number of subscribers to your podcast feed before you report direct downloads. Spotlight episodes that stand out in number of downloads. If possible, include player statistics, such as Spotify or your home site's embedded player. If your podcast is seeing great growth, include a chart that shows these increasing numbers. (For more about listener statistics, see the handy nearby sidebar, "Can you ever really know the size of your audience?")

>> **Display your show schedule.** If you post an episode several times a week or once a month, include a calendar or schedule of when your updates happen. If your show has no set schedule, be prepared to explain your episode release methodology. Many advertisers expect some sort of consistency (read: reliable exposure) from the places their ads are running.

>> **Provide demographic information.** Some advertisers want to know the general make-up of your audience. The more detailed you can get, the better. At a minimum, show a breakdown of gender, age range, and household income level. These statistics can be difficult to gather, but doing so increases your chances with many advertisers. Consider asking listeners to take a voluntary survey. If you ask your listeners to take a survey, it never hurts to offer them an incentive. Perhaps one lucky random survey taker wins a $50 Amazon gift card. There are many sites like http://surveymonkey.com, pictured in Figure 14-3, that cost a few dollars. Remember — sometimes you have to spend money to make money.

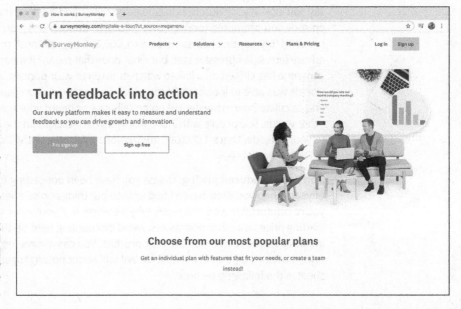

FIGURE 14-3: A service like SurveyMonkey allows you to find out more about your audience, something potential sponsors will want to know about your podcast.

- » **Showcase your popularity.** If you get 50 comments on each of your show note entries, talk about it. Technorati (`http://technorati.com`) and Google (`http://www.google.com`) can support any claims for the popularity of your site. Print any great testimonials showing your knowledge and expertise in the field. Favorable comments from other known experts go a long way, too!

- » **Provide a USB flash drive or online portfolio of your show.** With a press kit, you do want to make the best impression you can. That means you take the initiative and offer media that features your show. USB flash drives are the easiest media to use, with the drive showcasing only your best episodes (no more than five, even if you think your previous season was the best season ever!). Considering the compression format of your audio, you will not need flash drives larger than 500MB. This means money in your budget to get these drives branded with your show title and URL, if possible.

And in the "No, no, a thousand times no" corner, we have . . .

- » **Don't artificially inflate your statistics.** Grandmother always used to say, "Lies just cause you to tell bigger lies." Eventually you find yourself with an advertiser who wants you to explain how you came up with your numbers. Make sure you can support your claims. If they do not ask, you haven't dodged a bullet as the results your sponsors are expecting will not be what your results yield. Nothing good ever comes from fudging the numbers.

- » **Don't use terminology without really understanding what you're talking about.** Marketing and advertising professionals have a vocabulary all their own. ROI. Conversions. CTR. Maybe you do know what these mean; but if you don't, peppering your media kit with these terms is dangerous. For example, say you look at your numbers in Google Analytics and discover that an episode you promoted received over 1,000,000 impressions. You then proclaim "My podcast has earned over 1,000,000 impressions!" to potential advertisers. An impressive stat, but what does that mean? It means that someone has clicked on a link to watch or listen to your podcast, and that Google was able to look at the network of where that click originated from and calculate how many people potentially saw or heard your episode. In other words, 500 people with 2,000 followers each clicked on the link leading to your episode. That's 1,000,000 impressions. Welcome to the Doublespeak of Digital Marketing.

- » **Don't offer reduced pricing.** Unless you have been podcasting for a couple of years, then you likely haven't had time to put things on sale yet. Pick a price you're comfortable with and justify why it's worth it. If you're not sure of a starting price, ask other podcasters. Avoid discounting right off the bat; it implies that you're over-valuing your product. You can always negotiate the price, but don't try that in a media kit. We talk about putting together a rate sheet in the following section.

>> **Don't include a copy of your entire podcast season on a USB flash drive.** People like to know what they're getting themselves into. You have a great opportunity to offer that courtesy to potential advertisers — to let them listen enough to sample the wares. Remember, however, that their time is limited. You may include entire episodes if you like, but as mentioned earlier you may find better success by stitching together a "Best of . . ." reel from your show instead of a full season of your podcast. Select shows, or even highlights from your favorite episodes, take less time to consume.

>> **Don't make a halfhearted effort.** Okay, you don't have to create a full-blown presentation that would make the mavens of Madison Avenue envious of your skill set, but we're not talking Elmer's glue and crayons here, either. Treat it like a business; invest the right kind of time and money to make your media kit look (and sound) as professional as you can.

CAN YOU EVER REALLY KNOW THE SIZE OF YOUR AUDIENCE?

Tracking down listener statistics is elusive, and that's quite an understatement. Although many of the podcasting pundits will tell you how important understanding your audience size is (especially when you're trying to attract advertisers and sponsors), it's difficult for them to agree on the best way to snag those numbers.

Web server analysis tools are wholly unreliable in this regard; after all, they weren't designed for this purpose. They can tell you how many times your podcast episodes were requested, but not how many of those requests were successful; and if successful, how many of those downloads were listened to. They can tell you how many times your RSS 2.0 file was accessed, but they provide no tool to filter out requests made every five minutes by the same obsessed podcast app.

The most widely used model of determining audience size is the *bandwidth-division method*. This requires you, the podcaster, to know two pieces of data: the average size of your podcast media files and the total amount of data transferred in a given month. For example, suppose your ISP reported that you transferred 350GB of data in the month of May. Convert that to MB by multiplying by 1,024 (the numbers of megabytes in a gigabyte) to get 358,400MB of data transfer.

Okay, suppose that during that month, you put up four podcasts, each with a size of 10MB. Divide the total number of transfers (358,400) by the average size of your file (10MB) to get the "total" number of accesses to that file — in this case 35,840. Finally,

(continued)

(continued)

divide that number by the total number of podcast media files released (4, in this case). The answer (in theory, at least): You have approximately 8,960 listeners to your show.

This number, however, does not take into account the data transferred by your web server simply to run your website — say, images and text. Nor is it helpful when your podcast media files vary widely in size. And it still doesn't factor in those folks who downloaded 90 percent of your show before the connection broke.

That is not to say developers are not trying to come up with a better system. In June 2017, Apple introduced iTunes Connect (https://podcastsconnect.apple.com), a way to view analytics for listeners using its Podcasts app. Granted, this is not a complete picture of your audience, but it can provide some good insights to a sector of them, including trends, geographic locations, and average listening time per device. Of course, the analytics landscape is always changing long after this book hits the shelves — this is why we have *Podcasting For Dummies: The Companion Podcast* (http://podcastingfordummies.com), where we review and discuss current developments.

Our advice on this matter: Find a method that seems to work for you and stick with it. It's the best you can do . . . for now.

When you have your content figured out, go buy a nice, pretty binder. Or better yet, order some custom ones online that feature your podcast logo, slogan, and website address. You can get them online from VistaPrint (http://www.vistaprint.com) if you want large quantities. For smaller runs, see what your local office-supply or copy center can provide.

Establishing a rate sheet

A *rate sheet* is simply a table of what you charge for ads and other services like audio or graphic production related to advertising. And you thought you were headed into uncharted waters by just creating a podcast? How about figuring out a fair price to charge for running advertising? That's by no means a science. Heck, it's not even an art form at this stage! It's total guesswork — picking a number, throwing it out there, and seeing what sticks.

You can start by reviewing rates of other podcasts with similar topics or audience sizes. A rate sheet can be as simple as a fixed price for a given ad segment, but we like to think podcasting is more flexible than that. Can you offer cheaper prices for shorter ads? Can you offer prices for different placement within the show, keeping in mind that the closer to the top of the show, the more valuable the ad placement? You can offer decreasing rates based on how many shows carry an ad. How about adjusting the price if there's an ad in your show along with a link on your website? Are you capable of offering production services for an audio or video ad? All of these options can be combined in a variety of ways to complement each other.

Getting a Sponsor

Sponsorship and *advertising* are often used as interchangeable terms, but they're actually distinct approaches with different requirements. For the purposes of this discussion, we consider sponsorship as a relationship you (the podcaster) have worked out with an organization that regularly funds — and has a vested interest in the success of — your show. (Obviously, you can't say that about all advertisers.)

Sponsorships were popular — almost to the point of exclusivity — in the radio and television industries. During that time, corporations would actually create the various programs as an advertising vehicle — they could even censor content if (for example) some writer goofed and had the villain using the sponsor's product. Modern-day advertising deals, which allow a station or program to drop a 30-second ad into its already-established programming, came into play much later; these days, such deals dominate the scene.

REMEMBER

Understand that obtaining a true sponsor for your show likely means you give up some creative control. Instead of paying you a few dollars to run a pre-produced ad, your sponsoring corporation underwrites your entire show, or at least a significant portion of it. Rather than getting you to take a break from your content while you talk about its product for a few moments, your content may become *dedicated to* talking about its product.

Examples of sponsored podcasts include Mike Rowe's *The Way I Heard It* podcast (http://mikerowe.com/podcast/) and WNYC's *Note to Self* (http://www.wnyc.org/shows/notetoself) which have been sponsored by such companies as Blue Apron, Squarespace, and Zip Recruiter.

Traditional media sponsorships are extremely rare. Some businesses have launched affiliate programs where vendors generate a code that podcasters can offer to listeners for discounts or exclusive deals. Other vendors will generate a specific URL that your listeners can use. Each time a purchase is made with your unique URL, you receive a portion of the sale. *Affiliations,* like the ones represented in Figure 14-4 and featured on *Happy Hour from the Tower,* are a low-risk investment that benefit both vendor and podcaster. If no purchases are made, there is no loss on the side of the vendor. If the podcaster does generate revenue, the vendor is able to track these specific sales (use of the checkout code or URL); and a podcast's influence can be considered for possible traditional sponsorship deals.

If you're thinking about approaching a corporation to underwrite your show, you need to provide much of the same information necessary for securing advertisers, as discussed earlier. In addition, however, you need to demonstrate how your show can help bolster the success metrics of the corporation. "It's a cool show that your customers will love . . ." probably won't cut it.

FIGURE 14-4:
Some businesses like Gnarly Nerd Clothing (left) and Ritual Motion (right) offer *Affiliate Programs* that offer low-risk promotional programs that tends to benefit both podcasters and vendors.

MORE VALUABLE THAN MONEY

Tee and Chuck haven't gotten rich off podcasting. They do it for the love (see Chapter 16). The return on investment they have gotten from picking up a microphone and hitting record has gone far beyond cold hard cash. We know, this is starting to sound a little "hippy-dippy love fest," but give us a moment. Sure, we had aspirations at one time of getting a nice source of funding and perhaps a fleeting moment of quitting our day jobs and turning our hobby (or should we say *another* hobby) into a full-time paid gig, but the fates had other things in store for us.

Looking back on it, though, your humble authors were not expecting the pile of indirect benefits podcasting has led them to, and through. Thanks to podcasting, both Chuck and Tee have found some great career opportunities. As his gateway to social media, Tee not only wrote other books about social media but is a full-time professional in social media, managing social media platforms, and devising strategies that resonate with audiences around the world. Podcasting made Chuck a better presenter, which led to promotions at his day job, and even prepared him for an unexpected job transition in 2010. Audio and video productions are now a major part of Chuck's day job. With tens of thousands of followers and millions of video views, he's become a bona fide celebrity in his industry. As for Tee, his own understanding of podcasting and social media sent him coast-to-coast and around the world, leading workshops and seminars on the subject. Finally, let's not forget the awesome network of people and amazing friendships both Chuck and Tee have forged over a decade of firing up microphones and sharing their views through RSS feeds.

They cannot promise everyone future success such as they have been blessed with, but keep this in mind: When the people in your life question you for buying some (more) electronic gear, tell them you're doing it for more than money.

There's no secret sauce that irresistibly attracts sponsorship. Each business or corporation has distinct goals and objectives. Any podcaster trying to solicit sponsorship would be well served to understand these goals and objectives inside and out, and be prepared to clearly demonstrate how sponsoring a podcast can help the company achieve those goals.

Asking Your Listeners for Money

There is one group of potential financial backers out there who couldn't care less about your fancy media kit: your listeners. They couldn't give a hoot about your anticipated growth curve in subscribers or your ratio of direct downloaders to feed-based audience. They do have a vested interest in your continuing to produce the very best possible show, each and every episode — and they are (usually) content to let you take the show in the direction you feel it should go.

Sometimes your show is important enough to them that they're willing to pony up. That's why the following section discusses some novel ways you can go about soliciting your listeners for funds to help offset some of the costs of creating a podcast.

REMEMBER

Please don't turn your podcast into a weekly telethon. Most people tolerate the time when NPR goes into pledge-drive mode, but no one looks forward to it. If you're going to ask your audience for money, do it tactfully and as rarely as possible. Please.

Gathering listener donations with PayPal

Some of your audience will so love your show that they'll happily hand over some of their hard-earned money — if you'll only ask.

Asking for listeners' support is a two-step process: First you ask for the money, and then you provide an easy and convenient way for your listeners to send you money. PayPal (http://www.paypal.com) has been handling small and large web-based transactions for years. A PayPal donation link can be fully integrated into your website with minimum hassle. Here's how:

1. **Log in to PayPal.**

   If you don't have a PayPal account, click the Sign Up link and follow the simple instructions. You need a valid credit card to sign up.

REMEMBER

   PayPal, like much of the web, is subject to change. The flow of these steps might be slightly different at the time you're reading this. Regardless, the steps you need to follow will basically be the same.

2. **From your Account Overview page, click the Manage Buttons in the Selling Tools section on the right side.**

3. **Click the Create New Button option located in the Related Items section.**

4. **From the Choose a Button Type drop-down menu, select the Donations option and then designate a region, a language, and a button style; when done, click the Continue button.**

   You are taken to the Make Your Donation Button page, as shown in Figure 14-5. If you have a custom button already created, you can point PayPal to its URL, but you should keep it simple and use PayPal's design.

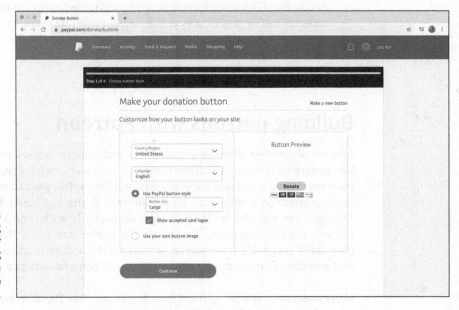

FIGURE 14-5: PayPal makes accepting donations as easy as selecting options from a series of drop menus.

5. **Add any additional information to your button. When done, click the Continue button.**

   We recommend using *"podcast name – Donations"* to help you identify your donations if you have other types of PayPal income.

6. **Set the Currency, default amount, and donation subscription options. When done, click the Continue button.**

7. **Review additional options for your Donations button.**

   These options include demographics, routing from PayPal to another site, and additional HTML variables.

TECHNICAL STUFF

   If you're familiar with using HTML forms, you can later add a suggested donation amount, and allow your listeners to change the amount if they desire. Get a copy of *Coding All-in-One For Dummies* (Wiley) by Nikhil Abraham if you need help with editing forms.

8. Click the Finish and Get Code button to create the code for your Donations button.

You have some decisions to make about the custom HTML code for your button.

9. Do some minor editing of HTML in order to incorporate your button into your podcast's blog.

Adding the HTML should simply be a copy and paste to your template or widget. If you need to go deeper into editing a template in order to accommodate your donation button, take a look at *WordPress All-in-One For Dummies,* 4th Edition by Lisa Sabin-Wilson.

This should do it for the donation link.

Building patrons with Patreon

Another alternative that many podcasters and creative artists are exploring for fan-generated revenue is Patreon (https://patreon.com). Patreon offers a couple ways for you to get paid for your content. Like PayPal, you can set up a monthly billing system to charge your loyal followers at the beginning of each month. However, you can also set it up to tap your patron's wallets on a "per creation" method. For the podcaster, this means the harder you work, the more you can potentially get paid. For the patron, it means if the podcaster starts to get complacent and doesn't produce any new content, you don't have to pay.

Many podcasters will set *up tiered patronage,* much like the different levels seen on streaming services like Twitch or YouTube, or in campaigns found on *Kickstarter* (http://www.kickstarter.com). For example, for $1 per episode, you get a special episode, show notes, or a postcard once a year. For $2 per episode, you get your name read on the show and a T-shirt. You see how it goes? If you support a show, you get goodies. Each podcast has its own take on what it charges and what it offers.

But how effective can Patreon be? For writers Chris Lester (http://www.patreon.com/authorchrislester) and Phil Rossi (http://www.patreon.com/philrossi), Patreon has proven to be quite the motivator in creating new fiction (see Figure 14-6). Chris Lester, host of the award-winning *Metamor City* podcast, had stepped away from storytelling for quite some time before returning with *The Raven and the Writing Desk,* his own author interview show that also journals his own return to writing fiction. His accompanying Patreon supported him writing on a consistent basis, even when finding himself unexpectedly in-between jobs. For Phil Rossi, author and podcaster behind *Crescent, Eden,* and *Harvey,* his greatest challenge for writing fiction was time. His Patreon income allows him to invest less in lucrative performance gigs as a musician so he can continue producing more fiction. Both Lester and Rossi offer their Patreon investors exclusive fiction both in audio and digital formats and additional incentives when print versions of their podcasts are released.

Before launching a Patreon campaign, have a plan. Think about what you can offer the potential contributor to make them want to sign up. Once you have your plan, create a Patreon account by following these steps:

1. **Click the Create on Patreon button, located in the upper right of the browser window.**

   You can use a Google or Facebook account to activate your account, or you can create an account using different contact information.

   If you are already set up with Patreon, simply click the Log In link to the left of the Create on Patreon button.

2. **Pick two categories where your podcast best fits and then click the Continue button.**

   Yes, there is a category specifically for podcasts!

3. **Honestly and accurately answer the question about your content by clicking the appropriate option.**

4. **Select the currency you want to use with your Patreon account and then click the Continue button.**

5. **Connect one of your social media platforms as a verification source to secure your custom Patreon URL and then click the Continue button.**

6. **Click the Start Customizing button and begin to put together your Patreon page.**

   Using a simple GUI, Patreon assists in building your website by having you enter basic information like your name, a profile picture, what you're creating, and appropriate social media links. You can even point Patreon to a short welcome video going even deeper into what you intend to create.

7. **When you are done with each section (and your checklist, as seen in Figure 14-7, is completed), click the Save Changes button.**

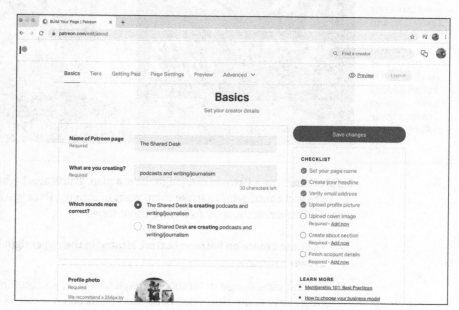

FIGURE 14-7: Patreon helps you create a website as well as set different levels of rewards for a variety of donations and how you will be getting paid by your fans.

8. **Click the Tiers option, located to the right of Basics, and then click the Customize button to edit and add tiers for your Patreon page. When done, click the Save Changes button.**

9. **Click the Getting Paid option to make changes and set your payment schedule. When done, click the Save Changes button.**

10. **Click the Page Settings option to make any changes you'd like to apply to your Patreon page. Click the Save Changes button.**

11. **Click the Preview option to review your Patreon page. If everything looks good, click the Launch button.**

    At any time, you can update and edit your Patreon page. Under the Advanced option, you have additions you can make to your page to offer your backers even more. Once you click the Launch button, your page is live and you are ready to go!

Now you can put a link or button on your website guiding your audience to your Patreon page, announce it on your show, or any other way to let people know how they can help you. Once a month, it automatically withdraws money from their account and deposits it in yours.

REMEMBER

Patreon takes a service fee for hosting and managing your content. Typically this is 5%. However, be sure to review the payment processing fees to understand how much you can expect from your donors.

Selling stuff

Some podcasters offer merchandise for sale as a way to support their show financially — T-shirts, hats, mugs, autographed pictures, you name it. If it's sellable, chances are good that some podcaster out there is selling it.

If you're contemplating selling merchandise via your podcast, you fall into one of two groups:

>> **You have merchandise to sell.** Musicians, authors, artists, and craftspeople fall into this category. For these podcasters, offering CDs, books, prints, or other items to the listening audience can bring in significant revenue. (Down the road, there's no reason why it shouldn't be possible to make a living via podcasting.)

Once again, PayPal (www.paypal.com) can be very helpful in taking away the technical hurdles for selling items online and integrating them into your podcast and website. If you can master the process for setting up online donations, you aren't far from having PayPal work as your entire online shopping cart.

>> **You need merchandise to sell.** If you need stuff to sell, CafePress (http://www.cafepress.com) is more than happy to step in and offer its assistance. Remember the T-shirts and other stuff mentioned earlier? You can go to the trouble of making those yourself, carrying an inventory, and shipping the orders out as they trickle in — or you can let CafePress take care of all that for you.

Setting up a CafePress storefront is a breeze, though it's a highly configurable task, and how you set it up depends on what you want to sell. A basic shop is free to set up and to use. Your listeners buy the T-shirts and coffee mugs (or whatever), with your logo or design on them from the CafePress store, and CafePress pays you a commission. It's very simple and requires only some quick setup information from you to get started. It takes about five minutes to set this up; CafePress takes care of the rest.

REMEMBER

Don't expect to make tons of cash via CafePress. Unless you can convince a large group of your listeners that they simply must have an $18 T-shirt with your logo on it, we wouldn't recommend quitting your day job just yet.

Is it possible to make money at podcasting? Sure, it's possible. You can find plenty of realistic success stories where people are turning their passion into a day job. It could be the podcast itself, suddenly becoming a runaway hit. It could be hosting a workshop or a class where you teach students how to podcast. It could be an opportunity where you produce podcasts for an organization, or a group of organizations. It could be getting together with someone you've known for years — a friend from the circles in which you podcast — and writing a book on how to podcast. There are a lot of pod-sibilities (oh, come on, your authors are both dads!) with this platform we are introducing you to, and there are plenty of options ahead. You can consider one offered in this chapter or a combination of them, all working to turn what you love into a day job.

So long as you remember the paycheck is not the sole reason you fire up the mics. Without the heart, the voice doesn't happen. But when it does, the paycheck is a nice perk at the end of an episode.

Chapter **15**

Podcasting for Publicity

"**T**rue" podcasters bristle at the growing commercialization of podcasting. With many corporate podcasts dominating Spotify, Apple Podcasts, and Stitcher, independent podcasters have voiced many concerns that boil down to one: A "true" podcast is done for passion, not for profit. The movement towards podcasting as a moneymaking venture goes against the grain of a "true" podcast. The words, opinions, and emotions should all be aimed at the listeners' hearts and minds, not their wallets.

Well, yeah, that's all fine and dandy — but before we cheer on the sentiments of the purist, indie podcaster, sheer love of the medium does not pay the bills. Also, when juggernaut organizations like NPR create runaway hits with millions of listeners, you won't hear their producers or creators thump their chests and proclaim proudly, "No, you go on and keep your commercial revenue and corporate bankrolls. We're going to keep on podcasting with our USB mic and preamp."

Truth be told, podcasting has proven itself to be a great way of reaching consumers. It's economical, unobtrusive, and can build a potential audience.

With blogging, social networks, and podcasts all vying for their place in the corporate world, all these platforms are finding themselves becoming integrated as part of a business's marketing and promotion machine; and with the right directors, strategists, and content producers in place, these platforms are gaining traction. Podcasting has come a long way since the first edition of this book, but still remains an unproven medium for many larger corporations. Because podcasting is audio-on-demand, there's still no easy way to measure numbers, feedback, and response in the same way radio and television advertisements are tracked. However, statistics from WordPress plug-ins like JetPack, online services like Google Analytics, and service providers like LibSyn continue to develop new methods to do so.

This doesn't mean a change in recognizing the potential of podcasting is taking its time. When iTunes officially recognized podcasts, its podcast subscriptions went from zero to 1 million in two days. Those kind of analytics did make the corporate world curious about the platform, realizing that some independent podcasts were discovering new audiences for their products (and broadening their existing audience) one MP3 at a time. If the end goal of your podcast is publicity, this chapter is for you. We go through several examples of where podcasting has paid off — maybe not in dollars and cents, but in good word-of-mouth buzz.

Podcasting and Politics

In the political arena, innovation can be a political candidate's best strategic option in winning an office. The "podcasting fad" that some of Capitol Hill's old guard snickered and scoffed at is now part of their own outreach initiatives. Individuals interested in holding office — be it local, state, or national — are sidestepping conventional media and adding podcasting to their platform.

REMEMBER

What we're defining in this section as a *political podcast* is a podcast hosted by a political figure or an individual seeking a political office. (We're *not* talking about podcasts where a host discusses current affairs, rants about the state of the world, or cracks jokes about the latest scandal.)

So, what can a podcast do for you, the tech-savvy politician?

>> Podcasting (and its companion blog, if you use a blog as part of your delivery) connects you directly with your constituencies.

>> Podcasting can reach young voters, the elusive demographic that can easily make or break a victory at the polls.

>> Podcasting avoids the bias that creeps up in media outlets, allowing you as a candidate to present agendas and intentions uninterrupted.

When going political with a podcast, whether you're running for the U.S. Senate or podcasting your term as School Board representative, here are a few things to keep in mind:

>> **Keep your podcast on a weekly schedule.** Monthly podcasts don't cut it; the tide of politics is in constant flux. Daily updates are less than practical, but weekly podcasts work well for ongoing political issues, and you can shape their content to remain timely enough to fit the podcasting medium. Set a day for delivering your weekly message and stick to that schedule.

>> **Focus on the issues, not the opinions.** A political podcaster will always walk a fine line between public servant and political commentator. You need to stay focused on the issues. If you suddenly start hammering away with opinion and commentary, you'd become more like Sean Hannity or Rachel Maddow. If you want politics as your podcast's subject matter, ask yourself whether you're looking to help listeners understand the issues accurately and take meaningful action, or whether you're just ranting and venting to entertain people who want their opinions reinforced.

>> **Give your listeners a plan for action.** When you cover the issues in your podcast, provide possible solutions to the pressing matters of your community and your constituents. Whether you're detailing blood drives or fundraisers, or launching an awareness campaign for cancer research, increase that divide between *political figure* and *political commentator* by offering listeners ways they can get involved in the community and make a difference.

WARNING

Podcasting is definitely an avenue to explore if you're venturing into the political arena, but this is not your sole means in reaching out to voters. You will need to campaign, of course. Get out on the road, shake hands, kiss babies, and the like. Unless you are recording said baby-kissing and hand-shaking, campaigning will mean your podcast goes on the occasional break. Make sure to communicate to your listeners that changes in the posting schedule are upcoming. When you can, give a podcast from the road to keep the communication lines open.

Anytime a podcast goes public on the issues, its creator walks a fine line between public servant and political commentator. The following podcasters all keep their content focused on the issues and less on what they "feel" are the issues.

The first politician to host a political podcast of this nature was North Carolina Senator John Edwards with the *One America Committee Podcast with John Edwards* back in 2005, covering poverty, Internet law, and environmental issues. Senator Edwards's wife, Elizabeth Edwards, was also featured, raising awareness on breast cancer. The success of the *One America* podcast caught the attention of the former govern-ator himself, Arnold Schwarzenegger, who podcasted his weekly radio address. Follow-up podcasts presented by politicians included the Oval Office with George W. Bush and Barack Obama (see Figure 15-1) hosting podcasts on a weekly basis, much in the same manner as FDR's Fireside Chats only this time happening on your mobile devices.

FIGURE 15-1: Barack Obama, at the time of this writing, carries the distinction of being the last sitting president to host a weekly podcast from the White House.

At the time of writing this fourth edition, shockingly, politician-hosted podcasts are something of a rarity. Perhaps *The New York Times* struck close to home in its 2017 article "Politicians Are Bad at Podcasting" (https://www.nytimes.com/2017/10/27/arts/politicians-are-bad-at-podcasting.html) when journalist Amanda Hess remarked, "The lawmaker podcast boom is just another way that our political news is becoming less accountable to the public and more personality driven. But that's not the only thing wrong with it. The podcasts are also boring."

Ouch.

Perhaps there aren't a multitude of podcasting politicians to choose from presently, but this is hardly a reason to dismiss using a podcast as part of your campaign outreach. If you're wanting to get in touch with the people (as a way to respond when voters complain about poor communication with their representatives or the candidates running for office), why not invite them into your kitchen, offer them a cup of virtual coffee, and ask them to relax a bit? Instead of dishing up prepared statements from professional speechwriters, you can offer voters (and worldwide listeners) impromptu, candid, sincere opinions on issues facing the country and the world. Podcasting can prove to be a promising platform, but remember a podcast of this nature can carry far more impact when it continues beyond the campaign.

Telling the World a Story, One Podcast at a Time

The whole point behind promotion — be it for books, film, or other forms of entertainment — is to win prospective target audiences (or build on existing ones) with something new or to give a different take on a familiar commodity. For a fraction of the cost of print advertisements and broadcast-media commercials, podcasting opens markets for your creative work — and can even start to get your name into an international market. Your podcast can not only offer a gateway into the unexplored corners of your new world, as Tee and his wife Pip Ballantine did with *Tales from the Archives* (http://www.ministryofpeculiaroccurrences.com), pictured in Figure 15-2, but these podcasts can tie back to the property you are promoting, enhancing the intellectual property (or IP) experience with sly references and clever crossovers. If you're an established presence in the writing market, or any entertainment field, the fans you have nurtured, with time, will not only eagerly support your podcast, but also introduce your MP3s to reader groups, friends, and enthusiasts of the subjects you're writing about.

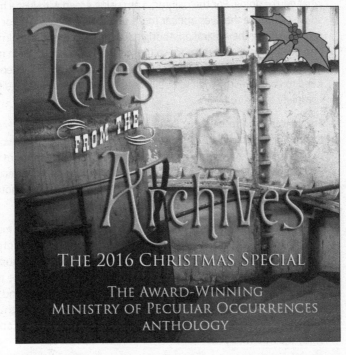

FIGURE 15-2: Tee Morris and Pip Ballantine introduce listeners to their steampunk world one short story at a time with their award-winning anthology series, *Tales from the Archives*.

Podcasting can introduce your original IP to audiences worldwide. For artists in more visual arts such as film, dance, painting, or sculpture, podcasting can serve as a journal leading up to the premiere of your work or a behind-the-scenes look at how works go from idea to fruition. It's an instant connection with your audience, and a great way to build an audience by getting them to know you on however intimate a level works best for you and your work. Planning a strategy for this kind of promotion only helps your agenda:

>> **When podcasting even in a visual media, briefly describe the action for the audience.** In *Masterpiece Studio* (http://www.pbs.org/wgbh/masterpiece/podcasts/), both hosts and guests add details about scenes and moments into their interview that accompany audio clips from productions being discussed. It is never taken for granted that people have seen the episode or special event this companion podcast is showcasing. (Many times, the host will give a quick spoiler warning.) In only a few words and a few seconds, *Masterpiece Studio* sets the scene for what's happening and why they are talking about it for the sake of listeners who are currently watching the latest PBS offering.

If you're documenting your visual art, it will take only a moment to describe what you're doing. For the painter: "I am using green with just a hint of black so we can make the eyes appear more unearthly, unnatural." For a dancer, "In reconstructing the Australian Aboriginal dance, you must remain grounded and deep in your squats, more so than what is normally seen in modern dance." Commentary like this, especially in a voiceover with video, does not need to go into every minute detail; but a few words are needed to create or complete a picture.

>> **For writers and musicians: Edit, edit, edit.** Awkward pauses, stammers, and stumbled words are obstacles for a writer introducing his work to the podosphere: They've gotta go! The approach is no different from that of an independent musician who podcasts a rehearsal session or a recording: You don't want off-key instruments and vocalists missing the high notes. Your podcasts need to sound sharp and clean.

Musicians, no matter if it takes 5 or 50 takes, should have their instruments in tune, lyrics clearly pronounced, and all notes sung on key. Writers should enunciate, speak clearly, and (most importantly) enjoy the manuscript. Each piece, whether music or printed word, should be a performance that serves as your audition to a worldwide audience. (While that may sound a bit nerve-wracking, don't think of it as walking out on stage so much as building something fine to send into the world.) Just have fun, and your audience will enjoy the ride with you.

If you need a refresher on the basics of editing, skip back to Chapter 8 for the primers on editing with Audacity and GarageBand.

» **Open or close your podcasts with a brief, off-script commentary.** Before beginning your latest installment or after your latest chapter concludes, a brief word from the author is a nice (and in many cases, welcoming) option for a podcast's audience. This commentary gives the author a unique opportunity to connect with readers, to share ideas about what inspired a scene, to promote an upcoming book-signing, or to give an update about what's happening in the next book's production.

This approach adds to the intimate experience of podcasting a novel, and podcast storytellers like Mur Lafferty, Phil Rossi, and Chris Lester use this to invite listeners into their real world after sharing the world they imagined. Let your audience members know who you are, and they'll show their appreciation through their support of your current and future works.

WHERE IT ALL BEGAN . . .

Allow us a personal account of the journey from podcasting fan to full-fledged podcaster. When *Podcasting For Dummies'* own Tee Morris was in the final rewrites of *Legacy of MOREVI,* a swashbuckling fantasy sequel to his debut novel, *MOREVI,* his publisher, anxious to know how Tee was planning to promote the novel, contacted him and said, "Okay, Tee, start thinking of neat ideas." Around this time, Evo Terra and Michael R. Mennenga had started podcasting their terrestrial radio show, *The Dragon Page.* The more Tee found out about podcasting, the more he pictured transforming *MOREVI* into a serialized audio adventure. As an actor who loved playing around with his Mac's audio and video features, he figured "Why not?" Reaching out to Mike and Evo with his plan, Tee set up his first rig and started recording.

The Prologue of *MOREVI: The Chronicles of Rafe & Askana* dropped into the Dragon Page's feed on January 21, 2005, and so began a completely new kind of promotion, different from what *The Dragon Page* had ever done before: a chapter presented every Friday, as read by the author. The feedback was instantaneous. Within weeks of Tee's first episode, two other authors stepped up and offered their works for podcast as well. One of them was Mark Jeffrey, author of the science fiction/fantasy young adult adventure, *The Pocket and The Pendant* (http://www.markjeffrey.net). The other author was adrenaline-hepped, in-your-face, über-intense conductor of mayhem, chaos, and carnage, Scott Sigler (http://www.scottsigler.com), and the book was *EarthCore*.

(continued)

(continued)

From here, the podcast novel — or *podiobook* — was born. Since then, authors both new and accomplished have turned to podcasting as a way of telling their stories with the world.

Podcasting continues to prove itself as a viable means of promotion for artists of any media unable to fund their own coast-to-coast tour. It may make conventional creators shake their heads at the idea of giving their works away for free, but the numbers and individual successes of authors taking this chance is the proof in this publicity. Now, before ink even hits the page, authors are building fan bases and getting their names out into the public.

And some of the really lucky ones get to write books for the *For Dummies* people!

Keeping Good Company: Community Podcasts

Slice-of-life podcasts that encourage community among listeners and fans are podcasts that *promote*. They can promote a show cancelled too soon into production, an issue affecting the well-being of a community, or offer a voice to a cause. The promotion comes from word-of-mouth advertising (*buzz*) that these podcasters generate from their thoughts, comments, and opinions on their subject — be it traveling across Spain, daily life in New York City, George Lucas's *Star Wars*, or Joss Whedon's *Firefly*.

Slice-of-life podcasts let the world into locations and clue people into possibilities that listeners may be curious about. After a few podcasts, you can even encourage listeners to experience that corner of the world, that idea, or join the community.

Do you have a cause you want to give attention to? Do you want to raise awareness in your county or district? Do you want to share the experience of preparing for a wedding or anniversary? Consider sound-seeing tour podcasts in order to build an online community through your podcast.

Creating a podcast to encourage testimony

Community is synonymous with podcasting. This book talks a lot about community, establishing a connection between you the podcaster and your audience, impatiently waiting for the next episode. Podcasts can also bring an existing community — a group with a shared interest, a community of homeowners, or a group dedicated to a cause — together and keep its members informed. The podcast can reach audiences in and outside of your community, sharing your interests and concerns with others, making your community even stronger in the long run.

The WDW Radio Show (Figure 15-3, found at http://www.wdwradio.com) has been a long-running podcast about planning the best trip to the "House of the Mouse," Walt Disney World. Hosted by Disney expert, author, speaker, and entrepreneur Lou Mongello, the *WDW Radio Show* has built an impressive community around his family-friendly podcast, featuring a blog, videos, and live broadcasts as part of his audio travel guide.

Travel planning is not the only topic of discussion on the *WDW Radio Show*. Lou also hosts interviews with representatives from the Walt Disney corporation, shares comments from listeners about previous episodes or queries about best travel tips, and many personal anecdotes on everything from a favorite amusement park ride to the best Disney vacation memories. What makes the *WDW Radio Show* more of a community than just another podcast is in how its podcast works to not only inform listeners on getting the most out of a visit to Walt Disney World, but also encourage listeners to share their own tips, ideas, and stories about the "best" way to experience Disney, Epcot, and other properties. This community, in turn, promotes the Walt Disney image through the best of methods: word-of-mouth.

The podosphere takes great pride in its sense of community, but the podcasts showcased here are set apart because the community is encouraged to take a more active role in the issues, concerns, and points of focus the podcast is centered around. More than chiming in with feedback, the community takes an active role in participating with increasing awareness over the podcast's main subject. In some cases, the producing of the podcast brings the community together, either through listener contributions, listeners directing the course of the show, or listeners coming together for a common cause.

Podcasting for fun (while promoting in the process)

When it comes to promotion, no one does a better job in promoting your business than your most passionate fans.

They work for free, set their own hours, and sing the praises (if you are lucky) of whatever it is you happen to be producing. Many times, these podcasts are nicknamed *fancasts*, but these are podcasts where consumers independently sit down around mics and talk about your business, be it an entertainment property, a product, or some sort of service.

But what, you may ask, is the line between a fancast and a podcast about your business?

Well, if you recall the sidebar from Chapter 5, *The Expanse* has *The Churn,* produced by SyFy Wire. *The Churn* is hosted by SyFy and features authors Ty Franck and Daniel Abraham as cohosts, who write collectively as James S.A. Corey, author of *The Expanse* novels. This makes the podcast an "official" production, and upon listening to it, you know without question that anything appearing on this show is firsthand knowledge of what is happening on the set, coming from the creators of the world, and shared from the actors bringing these characters to life.

That does not mean *The Churn* is the only podcast about this popular SyFy offering.

The Expanse Podcast: Tales from the Rocinante (http://solotalkmedia.com/category/the-expanse-podcast), is hosted by Solo Talk Media, a graphic designer from Ontario, Canada not affiliated in any way with SyFy or Universal Studios. Solo Talk Media (also known as Mark) is a fan of *The Expanse* and launched this podcast to share his love for the series. Along with show recaps, Mark reports news on *The Expanse* cast and crew and offers his own speculation on how things will unfold over upcoming episodes. Other shows like Solo Talk's include *Beltalowda* (https://baldmove.com/category/the-expanse/) and *Crash Couch* (https://crashcouch.podbean.com/), two podcasts independently produced from SyFy's own official podcasts. These are examples of fans who are sharing their appreciation for this science fiction series, serving as an unofficial street team for the production.

WARNING

When it comes to fans podcasting about a specific property or a generic theme, don't expect all the opinions coming from the podcast to be positive. If fans don't like a direction or a decision taken in a series, they will share it on their podcast. A fan's podcast could be considered the highest form of feedback, and should be regarded as such. You might like the podcasts supporting your favorite sports team, show, or organization. You also might hear some opinions radically different from your own.

Between fancasts hosted by experienced sports journalists (*Puck Soup* at https://pucksoup.libsyn.com/website) or passionate hockey fans like the hosts of *The OilersYYC* (http://oilersyyc.ca), the National Hockey League receives regular promotion and attention free of charge. Irish culture is also given plenty of attention through podcasts like historian Fin Dwyer's *Irish History Podcast* (https://play.acast.com/s/irishhistory) and celebrated musician Marc Gunn's *Irish and Celtic Music Podcast* (http://celticmusicpodcast.com), seen in Figure 15-4. What's terrific about these podcasts is they can actually work to not only promote your passion but promote your own brand. Marc Gunn, for example, in showcasing Irish and Celtic music, culture, and lore on his podcast, also spreads awareness of his own brand as an accomplished musician. The podcast, blog, and companion app all offer listeners a chance to find his music alongside the music of other independent musicians featured on his podcast.

Community-driven podcasts cover a wide range of audiences. However, all communities share a similar mindset, and you can apply these sound production (see what we did there?) principles:

>> **You are the host, but it's not all about you.** Community podcasts should be about the community. Yes, there is room for personal thoughts and commentary, but in small doses.

The podcast is about the community and how it interacts with the world around you; that is what the content should focus on. Your podcast can feature other members of the community who share the same opinions as yours or even take opposing viewpoints (a spirited debate can up your show). Just remember that the community-based podcast is not about you personally, but about how you see the world, how that connects with the people around you in the community, and how all that comes together in the pursuit of a common interest.

>> **Avoid the negative.** It would be easy to turn a podcast into a gossip column or a personal rant against the very concept that brought the community into being. While there is no law or ethic barring you from speaking out or voicing concerns, a community is based on support. Whether you consider yourself a fan of Harry Potter, Apple Computers, or your local county, your goal in a community podcast is to remain positive and celebrate the benefits of being part of the cooperative spirit. If there is a matter of concern in your community, then there's room for debate and action, as with a political podcast — offer some possible solutions to these issues.

Regardless of the kind of community you're chronicling, your podcast should work much like glue — helping to keep supporters together in the face of problems (instead of just crying in your collective beer) and celebrating what gives them joy. Reinforce that sense of community and keep your podcast strong.

IN THIS CHAPTER

» Connecting with the passion of
   the cast

» Asking, "Why do we do it?"

» Keeping the passion alive in your own
   podcast

Chapter **16**

Podcasting for Passion

O ur mission from the beginning of this book has been to fill you in on what a podcast is, demystify the technical aspects of getting your voice heard on media players around the world, and assure you that those with the drive to make their message happen can do it, with the right tools. But a podcast is empty if you don't have that drive, that passion to get your podcast up and running and keep it running, whether the episode goes out every week, every other week, or every month. We drop that word a lot throughout the book: *passion*. What do we mean by "podcasting with passion"?

Okay, if you take the dictionary as a starting point, *passion* means "a keen interest in a particular subject or activity, or the object of somebody's intense interest or enthusiasm." You can have state-of-the-art audio equipment, hire the most engaging vocal talent, and employ the best engineers, but that isn't what makes a podcast a podcast. The artists and creativity behind a podcast make an investment — not simply financial or physical — and collect the windfall in listener feedback and download stats.

Yes, a podcast can make money, promote a cause, or bring attention to social issues, but without that passion, you might as well sit behind a microphone and read the instruction manual to your mobile phone. Passion is what motivates relative unknowns to slip on the headphones, step up to the microphone, and let their words fly.

This chapter won't presume to teach you how to create, cultivate, or conjure your own passion. It's something you have, or you don't, for a given topic — and we suspect everybody has it for *some* topics. Instead, we offer some real-world examples of how others apply their passions to podcasts — and we tell you about some tools to apply a little passion-power of your own.

The Philosophical Question for All Podcasters: Why Do We Do It?

It's going to come up sooner or later: Someone will question your motives, rationale, and even sanity at your investment of so much time and energy into a podcast. We highly recommend deep nonsensical answers such as "It's because of the cheese . . ." be returned to these people. Then watch their eyes start to twitch. That never stops being funny.

The following sections tell you some of the ways you can take that passion and channel it in to a truly effective force behind your podcast.

Gaining perspective on passion

"Get some perspective!" they say. Well, we agree with them (whoever they are). As a podcaster, you need a firm idea of what your show is about. If you're not sure what your point is, your audience isn't going to be sure, either!

Marc Blackburn's *America at War* podcast (Figure 16-1 at http://www.america atwarpodcast.com/) started in February 2016, relatively recent in geologic terms. The content, on the other hand, has been centuries in the making and has taken decades of Marc's life to collect and document his particular story about military history. Marc, a Ranger for the National Park Service, has a PhD in American Military history and has been in love with the topic since he was a teen — let's just say a "long time ago." He loves his topic and he loves to write. That's what we call a certifiable subject matter expert!

Marc was one of a dozen or so people who participated in a podcsating course offered to the National Park Service western region by Chuck Tomasi in 2009. One day he found himself with training, the material, and enough Amazon gift cards to get some equipment so he decided it was time to let the world know about his love for military history.

He has a well-thought-out method to his madness that can serve aspiring podcasters well:

>> **Learn the landscape.** Take time to find out what shows on your topic exist. Like any good prospective podcaster, Marc did some listening to other similar podcasts to understand what is out there and identify a unique place he can fill in the podcast universe.

>> **Look for a unique angle.** The last things your listeners need are carbon copies of other podcasts or (perish the thought) radio shows based on a template. If your passion is commonplace, focus on a niche or small aspect that you have the most knowledge about. Many of the shows Marc heard focused on specific events, often leaving the listener desiring more context of how we got those events. Marc opted for a chronological approach starting with the settlement of Jamestown in 1607. Until Chuck listened, he wasn't aware that the reasons for our early colonists seceding from England go back so far.

>> **Make it personal.** Let's face it, history in high school was pretty dull because it was mainly facts about people and places a long time ago — okay, as teens we also had other things on our minds to distract us, too. Marc explains that it helps to make the listener feel a part of the story. He's got expertise in this area as well having interpretation training from the National Park Service. Here's wishing we had someone like Marc leading our high school history class. It would have been a lot easier to relate to what these people went through since we were of the nerdier persuasion who didn't have to think about a date on Friday night.

FIGURE 16-1:
Marc Blackburn puts his PhD in Military History to good use on his show *America at War*.

Podcasting passion with a purpose

Podcasting with a purpose is an essential. By doing so, you have a destination. It doesn't matter if that destination is to educate, showcase your material, or entertain. Pick something as your core reason for speaking and don't be afraid to have fun with it and make it exciting.

That's what Chuck has done with one of his productions for his day job. Twice a week, Chuck creates a video for developers inspired by questions and comments he reads in the online community. The purpose is to educate and enable his audience to be more effective developers. The genesis was that he had been spending time in the community for several years answering questions and building a reputation when the idea struck him to start providing the thought process behind the typed answers, so he started a series.

His approach to podcasting can be summed up like this:

>> **A new perspective on content:** Chuck is constantly evaluating the good, bad, and ugly from his work projects and asking, "Would this be good in the show," "I should think about adding this in a future episode," and "Do I think people on the show will find this helpful?"

>> **Improvement:** It's hard to listen to another show, watch TV or a movie without thinking about the production, editing, lighting, vocal inflections, really everything that goes in to it and taking inspiration to try to improve his own product. That's called being professionally aware.

>> **Continuity:** Produce episodes on a regular basis with a consistent format. Chuck got a tip from a former podcaster a few years ago. She said, "If you want to make changes to your show, do it how you would cook a frog. You don't just throw a live frog in a frying pan or it will hop out. You put it in a pot of water and slowly turn up the heat." The metaphor, gross as it may seem, is also appropriate to podcasting. If you suddenly shift topics or format, your audience may think they got the wrong show.

>> **Because it's fun:** Sure, the primary purpose of the show is to drive interest with current and potential developers, but it's the passion for what he does that keeps the show going.

Sharing your passion with friends

Enthusiasm can be infectious. Podcasting can sometimes feel like a very solitary endeavor when it's just one voice behind the mic, but as we mention in Chapter 5, getting multiple hosts around the microphone can be a challenge. However, when the mics go live and the banter begins it can be a challenge to stop talking. It's that spark that all podcasters strive for, and sometimes, with the right people, the chemistry is instantaneous.

This shared passion was the driving force behind Tee's decision to call on his friends while working on the third edition of this book to launch a new podcast, *Happy Hour from the Tower* (http://www.happyhourfromthetower.com) seen in Figure 16-2. There are video gamers that are forces of nature dominating in *Destiny* like iLulu (http://twitch.tv/ilulu), Gladd (http://twitch.tv/gladd), Gigz (http://twitch.tv/gigz), and Red Queen (http://twitch.tv/redqueen). Then there's Nick Kelly, his son Brandon, and Tee. They may not be the best players in Bungie's epic space opera, but they can't get enough of the game, the lore behind it, and their own experiences within it. Every two weeks, it is that love of the game that brings them together on the mics to talk about the game. On occasion, they are joined by other streamers, voice talent, and even people who helped create the game. *Happy Hour from the Tower* is Tee's celebration and unabashed love for this science fiction epic, and he's inviting friends over to his studio both physically and virtually to join him in that celebration.

FIGURE 16-2:
Happy Hour from the Tower features three generations of console gamers that get on mic for the love of the game.

When you invite friends together to launch a podcast, here are a few tips to keep in mind:

>> **Make sure to have an agenda for the night.** Coming up with a plan is more than just saying to the people you work with, "We should do a podcast because it's cool. Here's what we talk about in Episode One." A podcast's

agenda, or game plan, or to-do list, should be an idea of what you will want to gather around the mics to riff about for at least five shows. If you have ideas that go well beyond Episode 10, then you might just be on to something. When getting the cast and crew together for your podcast, you will want to make sure it is clear what the topic of conversation is. Otherwise, you might find yourself recording awkward silence, and that makes for terrible audio.

» **Make the most of your time together.** If the subject matter is not time-sensitive, or if you have mapped out evergreen topics (subjects of discussion that are relevant no matter when they drop in a show's schedule) for a few episodes, then record these episodes in one night. Granted the ability to create a buffer for your podcast will depend on the running time for your episodes. If at all possible, record several episodes in case schedule conflicts occur.

» **Nominate one person to take the point in a conversation.** Usually in situations where many hosts are on mic, there should be one person calling the plays and prompting others for commentary. In a quick-paced, back-and-forth conversation, that might mean the lead host has to sit back and moderate the episode. What matters is that you stay on topic lest your episode goes bouncing off the rails and into the great abyss. Having a lead host — and that position can vary from show to show, depending on the topic — will keep you, your cast, and your episode on track.

» **Invite others to join in on the fun.** While there should be core hosts in every multi-mic podcast, inviting people to join in only adds to the dynamics of a podcast. It could be a formal interview, or better still, it can be an informal sit-down where everyone contributes. These sort of roundtable discussions may open your podcast up for more hosts, or some hosts may offer to pick up the mantle for a spinoff podcast out of loyalty to and joy from the podcast. We've seen this happen too many times to count. Open mics also offer different perspectives and backgrounds, providing your podcast a broader scope and reach.

A passionate love for the podcast

All successful podcasters share in common a true love for the podcast. Something about getting behind the mic and producing quality content is satisfying. That passion can tax you, especially if the podcast is an emotional ride; and if a podcast reaches a conclusion, it would make sense if the podcast's host decided to step away from the mic for a spell.

However, when you are one of podcasting's original voices, it is hard not to want to podcast, especially when embarking on new adventures overseas.

Such is the story of Evo Terra, a name very familiar in podcasting circles. Evo was one of the original advocates of podcasting, hosting alongside Michael R. Mennenga

The Dragon Page in October 2004. He launched Podiobooks.com with Chris Miller and Tee Morris, and joined Tee on the first edition of *Podcasting For Dummies*. "One thing we had going for us," Evo recalls, "is that Mike and I were already doing *The Dragon Page* as a syndicated radio show, both on Internet Radio and terrestrial radio since 2002, so we hit the ground running while many other podcasters at that time were figuring things out as they went. Many of the technical challenges of podcasting gear we already had figured out."

Evo loved podcasting so much, he brought in his wife, Sheila Unwin, when *The Dragon Page* launched a Y.A. Fiction spinoff series, *The Dragon Page with Class*. Together, Evo and Sheila podcast on *Evo at 11* and *The Opportunistic Travelers* which evolved into *This One Time. . . .* All three podcasts are very different, but all three are very much full of the intellectual snark that Evo and Sheila are known for. "When I was asked to deliver the keynote at the very first Podcast Movement event, I was told 'You should be podcasting more. . .' but by then Sheila and I were transitioning to leave the country," Evo recalls. "I had sold most of my equipment, but I had kept my H4n, a few cables, and a pair of Shure SM58s. It occurred to me that we needed to produce a podcast to explain what we were doing. I had been listening to *The Startup Podcast* with Alex Bloomberg, and I really liked that journalistic style. So during our drive out to California, I broke out the mic and started asking Sheila questions, and that was the first episode of the *Opportunistic Travelers*." As seen in Figure 16-3, *The Opportunistic Travelers*, while a dramatic change in the kind of podcast Evo was known for, still retains his signature of a light-hearted look at life and the world around him.

FIGURE 16-3: Evo Terra and Sheila Dee podcast their adventures around the world as *The Opportunistic Travelers*.

Sheila's own love of the podcast took root not on the technical side, but more on the content provider side. "When I first started, I was primarily voice talent, and it was a lot of fun. Initially, *The Dragon Page with Class* was an interview show, and I really loved the back stories of these books I've been reading and how they came to be." After *Evo at 11*, Sheila believed her podcasting chapter had concluded. As she states, "When we were getting ready to leave the country, Evo said, 'We'll update people in a blog format,' and I came back with 'That doesn't make any sense. You're a podcaster. We need to podcast this.' So we started *The Opportunistic Traveler*." Not only has podcasting made Sheila a better public speaker and presenter, Sheila gives podcasting credit for the close friendships she has cultivated over the years. "I have friends all over the world, and I've met people at different conferences all of whom know me from my appearances on podcasts."

What keeps both Evo and Sheila podcasting today, on a more personal level, is their online dynamic. "Our podcasts are our personalities amped up," Sheila says. Evo agrees. "It's a project and a platform that we not only can work on together, we work on it well together."

And it doesn't hurt Evo's passion at all that it has now become a source of income. Since returning to the States, Evo has started a strategic consulting service and helps others launch and maintain their podcasts, allowing them to focus on the content. And yes, he's still podcasting with a podcast about podcasting called *Podcast Pontifications* (https://podcastpontifications.com).

The desire to continue recording, producing, and uploading shows simply for the love of the podcast is not uncommon. Many podcasts are created out of that need. These shows and their hosts stand out in their dedication, infusing raw enthusiasm for this medium into every new show. It is this passion that brings the podcaster back to the microphone and invites new podcasters (like you!) to take the host's chair.

Holding Interest: Keeping a Podcast's Passion Alive

You have your podcast underway. You're planning to have a weekly show. The first month in, you feel strong, confident. But this is all in the first four episodes. How do you keep the momentum going?

In the early days of a podcast, you can easily see yourself continuing banter a year from launch date — but remember that schedule you planned? Have you taken into account sick days? Vacation? The occasional stumbling block of inertia like the "Do I really want to do a podcast today?" question. What about unexpected technical issues? Even the best podcasters need to step away from their mics, recharge their batteries, and then jump back into their recording. Personal health, well-being, and time to edit episodes if your episodes need editing (and if you're human, sometimes they do) are factors you also must consider. Add in dealing with conditions such as background noise and having to tax the strength of your voice, and podcasting can become less of a joy and more like a chore or a second day job. Plenty of podcasts begin strong out of the box, only to have their feeds go silent, and remain so.

Even with passion, momentum is difficult to sustain. How do podcasters keep that spark alive? Well — as with the answers to "Why do we do it?" — podcasters have a myriad of reasons. Each show applies different tactics to keep each new episode fresh; even amid diverse topics, you can find common threads between all podcasts that provide the momentum to forge ahead 50 episodes later.

Podcasting on puree: Mixing it up

After you've racked up a few episodes of your podcast, take a look at its format. How do you have it set up? Is it all commentary? What can you do to vary the content?

Just because this is your first, second, third — or seventeenth, eighteenth, or nineteenth — episode, that doesn't mean you've set your show in stone. (That's where you find the fossils of all sorts of creatures that didn't manage to evolve.) In this section, we look at ways to dodge that asteroid and keep your show from suffering an extinction-level event.

Don't be afraid to try different things with your podcast's format. For example, you might

>> Talk about a product, a service, or an idea that's only loosely related to your normal focus.

>> Interview a guest with a unique perspective on the focus of your show.

>> Experiment with adding a cohost, even if only temporarily.

>> Podcast from a remote location or from a studio if you normally do remote shows.

Technorama, hosted by one of your illustrious authors, has been in a near-constant state of evolution since inception in May 2005. It started with a single host and acquired a second after only a handful of episodes. Things stayed constant for a while, and then its hosts — Chuck Tomasi and Kreg Steppe — added an "On This

Day in History" segment to the show to review tech, science, and geek events that happened in history on the day the show is released.

With the addition of "On This Day in History" segments, Chuck and Kreg started experimenting with production elements and other kinds of regularly occurring features in the podcast. Soon the show, described as looking at the lighter side of technology, started covering odd news in the "Hacks and Strange Stories" segment and began scratching their heads at how silly people can get over tech with the "What the Chuck?!" segment. Following the wacky wake-up morning show format, *Technorama* gained a loyal following.

Fast-forward through the years (time travel saves belaboring the point), and the award-winning show continues to offer something different than its normal fare. While taking pride in their lighter look at tech, *Technorama* has also featured interviews with hosts of TV's *Mythbusters*, oceanographer and *Titanic*-discoverer Dr. Bob Ballard, and renowned physicist Dr. Michio Kaku. The guys have also ventured in to video podcasting, on-location recordings, and elaborate audio spoofs ranging from "How Can You Tell He's a Geek" (inspired by Monty Python's "How You Can Tell She's a Witch" sketch) and several parodies inspired by *Star Trek*, *Star Wars*, classic TV shows and more. Even today, Chuck and Kreg continue to challenge the status quo and try new things.

Your show may not have to go to elaborate lengths to stay fresh, but it does need to heed the creed: *Evolve or die.* Remember the story about cooking a frog mentioned earlier in this chapter — do changes like this slowly and get feedback from your audience to see what works.

WARNING

If you decide your podcast is going to be in a constant state of flux, keep in mind that the risk of alienating your listeners is ever present. Podcasting for podcasting's sake is like improvisational comedy: It's a skill. You can keep your passion alive by continuously changing the format of your show, but you run the gambit of making your show an acquired taste.

Starting a second (or third) podcast

Keeping the passion alive and well in one podcast can be found in variety, but what happens when you are finding yourself cutting off the microphones and still having a lot to say. Some podcasters decide to fire up the mics again and keep going. Sure, the occasional "super-sized" show isn't a bad thing, especially if the topic is timely and your normal running time can't cover everything you want or need to. What happens, though, when your special super-sized shows are becoming less and less special, but becoming the norm? What if your broad focus on a subject matter is becoming more and more focused on one aspect of it? Are you willing to risk overloading your podcast's audience with new content, and possibly burn out you, your cohosts, and your audience?

PODCASTING IS HARD

While updating the final chapters for this fourth edition, we went back and counted how many podcasts we had cited in the previous edition and checked how many were still producing content. We were shocked to find that over half had *pod-faded* — even more if you count the spinoff podcasts that the producers have created. We were also not surprised.

It is not uncommon to meet people at events who say "I'm a podcaster . . .," but look up their podcast and you find four episodes posted two years ago. Many podcasts launch with good intentions only to fall off the rails within five episodes. On the other end of the spectrum, you could find runaway success with your podcast and watch it grow into some kind of monster, to quote the front man for Metallica. This may mean that you find a good show to end on and then post that episode where you thank your listeners and then sign off one last time. Podcasting may get to be a lot of work. Editing offers its own challenges. Promoting and cultivating a podcast can be difficult. At some point, you are going to find a variety of reasons why you won't create a new episode. If you need to take a break, we get it. Life happens. Reflect on your passion. Is it still a driving force, or have other priorities taken over? Does your podcast still have a purpose for you (and your audience)? Is it still fun? If you aren't enjoying the podcast, how can you expect your audience to enjoy it? While podcasting is easy to do, podcasting isn't easy.

It might be time for another podcast.

Okay, that may double your workload. Well, no, it *will* double your workload; but spinoff podcasts allow you to follow this new rabbit hole you've discovered while keeping the original podcast's voice true. A new podcast also keeps your perspectives fresh and excited about what unknown discussions you have yet to record for upcoming podcasts.

If your gut instinct is telling you to begin a second podcast, go for it. Variety, not only in a podcast's original content but also between the different podcasts you create, can benefit both shows. Now, instead of the same old podcast every week, you can allow yourself different avenues to explore — and maybe in the new venture, attract a whole new audience.

When cybersecurity professional, cyberpunk author, and rock musician Nick Kelly first sat down in 2013 with Tee and Pip at an author event, he had no idea what new creative outlet would unfold before him. After a second in-studio visit on *The Shared Desk*, pictured in Figure 16-4, Nick took his first step in podcasting. "I had most of what I needed at home and I knew one of the writers of *Podcasting For Dummies*, so I figured I was covered," Nick says. "And I aimed to keep it simple: Get around the mics with the family and just geek out a little."

FIGURE 16-4:
Nick Kelly (right) joined his wife, Dr. Stacia Kelly (center) on an episode of *The Shared Desk* and unexpectedly stepped into a new creative outlet — podcasting.

Since its inception in 2014, the Geek Wolfpack brand, expanded quickly, all its podcasts carrying the same signature wit and wisdom from Nick, Stacia, and Brandon:

>> **The Geek Wolfpack Podcast** (http://www.geekfamilypodcast.com): The Geek Wolfpack is the Kelly Family at their nerdiest. Nick and Stacia are both science-fiction writers, while Brandon is an avid gamer. Together, these three settle around the mics to talk about their passions as a family in their travels, what they are watching on various streaming services, and what new app they are currently mastering on their smartphones. This podcast is about keeping it geeky in the family and the fun to be had when you do.

>> **ADHD D&D** (https://adhddnd.com): Part Dungeons & Dragons playthrough, part comedy, *ADHD D&D* features the adventures of a merry band of adventurers comprised of the Kelly family, the Morris family, a Dungeon Master and his wife in St. Louis, and finally, Steve "The Blind Gamer" Saylor (https://stevesaylor.net), an avid gamer and advocate for accessibility in gaming. Between the dropping of pop culture references to overall observations of how many ways the DM is trying to kill them all, *ADHD D&D* is less about the skill and prowess of our heroes and more about whether they will make it through a battle without quoting *Clue* or *Caddyshack*. This isn't Dungeons & Dragons unplugged. This is D&D unhinged.

>> **ADHD D&D Classic** (also https://adhddnd.com): Before the arrival of Steve "The Blind Gamer" Saylor, the wrath of Arrow Storm, and Ashimei the Teenage Barbarian adopted (with love and brute force) a dinosaur named George, Nick played a half-dragon War Mage who constantly gave the party's Ranger serious side eye over a pair of dragon hide boots. This was also the time of Shade's evolution into "The Murder Hobbit" and *"Protect the Healer!"* became the party motto. *ADHD D&D Classic* are the hours of content from the early days of their party's adventures that never made it on the *Geek Wolfpack Podcast*.

"When it got harder and harder to take three hours of what was some hysterical content and edit it to a ten-minute segment for *Geek Wolfpack Podcast*," Nick states, "I knew I was going to be making a spinoff podcast. Then, as I started editing segments and looked ahead at the hours on hours of content I had recorded, I knew it would be years before I would introduce Steve Saylor's blind wizard." In this interview, he admits an inconvenient truth: "Tee warned me. He did."

And while Nick manages the Geek Wolfpack Network of three shows, he also appears with Brandon on Tee's earlier-cited podcast, *Happy Hour from the Tower*. Considering Nick didn't even know what a podcast was when he and Tee met, this has been an epic journey for him.

If you're thinking about branching out with a new show, consider these tips:

>> **Take your audience with you.** If you're podcasting about triathlons, yet hold a secret passion for seventh-century Gaelic text, expect only a handful of folks to subscribe to both shows. However, if you move from triathlons to a podcast on sports medicine, they may find the new topic an easier pill to swallow.

>> **Cross-promote.** Even if your podcasts are quite close in scope, there will be listeners to one who aren't aware of the other. Although we caution against turning one podcast into a giant commercial for the other, plug the heck out of your other show. Within reason.

>> **Look for the niche within the niche.** If your main topic is broad, break it out into chunks and see what needs further exploration. Granted, a deep dive into the minutiae of a niche topic may reduce the size of your listening audience. But so what? This is about exploring your passion, not about gaining market share. And (we promise) there's always someone else out there who will want you to delve even deeper into the obscure topic you've just built an entire podcast around.

>> **Keep trying.** Don't expect to hit your stride in the new show by Episode 2. You have to try different formats, flavors, and ideas, just the way you did to make your first podcast perfect. (Or is it? Maybe it's getting a little stale. Time to spruce that one up, too. See the following section.)

When enjoying the journey of your first podcast, keep in mind that your next podcast may be closer than you think. Keep a close eye on your feedback; look for recurring themes that your audience likes to hear. If it's a broad enough topic and you can speak on it with authority, you may be able to craft a show around audience suggestions instead of having to sweat bullets to come up with something brand new every time.

Moving forward with a plan

If you are keeping the fires burning for your podcast whether it's a spinoff podcast, something new, or a complete rebrand of what you've been doing for some time, it's a good idea to move forward with a plan. So here are a few dos and don'ts — with the don'ts first.

WHO'S UP FOR A CHALLENGE?

Kreg Steppe, Chuck's cohost on *Technorama*, came up with a great idea in 2012. He challenged podcasters to publish a podcast a day for the month of August. He called his challenge "The Dog Days of Podcasting" (http://dogdaysofpodcasting.com). His idea was to get back to the roots of podcasting (late 2004 and early 2005), theoretically where there were so few shows that all the podcasters seemed to be listening to each other's show and had cross-show conversations with each other. Something said on one show would get a response on another show. The first year or two of DDoP there were only a half dozen or so podcasters who joined in, making it easy to subscribe to all of them and keep up with each thread. However, in recent years, the group has grown and listening to every single episode of all the shows has become something of a challenge in itself. Some of the nostalgia now is hearing from some of podcasting's original voices (see Chapter 20), but new voices also appear in Steppe's challenge. In fact, some people like Amy Bowen and Charlotte Kennedy have used the Dog Days challenge to launch their own podcasts, proving that you don't need a decade of experience nor lots of technology to participate in Dog Days. Kreg has made it easy to listen to all the podcasts without having to subscribe to each show individually with a master feed including all episodes for all shows during the annual challenge.

So if you want to see what a daily podcasting challenge would be like, give DDoP a listen. It truly is a time capsule of podcasting, created every year, for you to experience.

What NOT to do (Ack! Run away! Run away!):

>> **Don't use shock-radio techniques.** Hearing a podcaster drop a few words that would make most grandmothers blush isn't anything new, and it's a cheap way to get a laugh and attention. We personally don't have any moral compunction against the more colorful aspects of the English language. If that's your style, don't change it; otherwise, try to use them sparingly, okay?

>> **Don't swing between kid-friendly and adults-only.** If you're unhappy being a family show, give some advance notice so folks can unsubscribe. Hell hath no fury like a mother scorned (or even unpleasantly surprised).

>> **Don't get angry when you get dissenting email.** Not everyone will be happy with your decision to change things around or to add a new segment. Deal with it and move on. You can't please everyone — or "correct" their opinions — so don't even try.

What to DO with all your might:

>> **Cast your net wide.** Be open to a variety of new ideas and concepts and don't be afraid to try them out on your audience. Your loyal listeners will likely forgive you a few miscues along the way.

>> **Have a purpose for the change or addition.** Find the common connection with what your show has focused on before and talk to your audience about it. You can do this before or after the "newness" goes in the show, but your listeners will enjoy knowing the method behind your madness.

>> **Encourage feedback about the change.** When you do this, ask more than "Do you like it?" Find out whether your audience members feel differently about your show, if they think it's "fresher" or perhaps more appealing — and ask why. The changes you put in place are designed to give your show a boost; be sure and ask for confirmation that it actually happened!

REMEMBER

If you find yourself in need of a break, there's nothing wrong with taking some time out. As a courtesy to those who are listening, however — whether it's 20 or 20,000 — let them know you're holding off on new episodes for a few weeks so you can reorganize priorities, goals, and the whole momentum of your podcast in order to make it even more rewarding for them to listen to. The same courtesy applies when you've had an unexpected illness: Give your listenership a quick update on what happened; reassure them that your voice is back and that your podcast is back. As passionate as you may be about a podcast, always remember your health and your voice must come first. There's nothing wrong with missing a week or two of a podcast while you get back up to par.

When Podcasting, Be Like Bruce Lee

To me, ultimately, martial art means honestly expressing yourself. Now it is very difficult to do. I mean it is easy for me to put on a show and be cocky and be flooded with a cocky feeling and then feel, then, like pretty cool . . . and be blinded by it. Or I can show you some really fancy movement. But, to express oneself honestly, not lying to oneself — and to express myself honestly — that, my friend, is very hard to do.

— BRUCE LEE, FROM AN INTERVIEW ON *THE PIERRE BURTON SHOW*

When Tee was writing the first draft of this particular chapter back in 2006, *Bruce Lee: The Way of the Warrior* had been playing in the background. This happens to be a (really cool) documentary featuring recently discovered footage from Bruce Lee's then-work-in-progress *Game of Death,* rare commentary from The Dragon himself, and interview footage. In one of the interviews came the preceding quote — a compelling moment when Bruce Lee sums up, in his own words, what the martial arts mean from his perspective. As Tee pondered those words, it struck him that Bruce was basically saying Martial arts are all about keeping it real.

And (for whatever reason) he blurted out loud, "If Bruce Lee were alive today, he would have a podcast!"

Part of the passion in a podcast is cultivating honesty with yourself and your audience, and that honesty in your podcast is what keeps listeners coming back. "To thine own self be true . . ." — one of Shakespeare's most-quoted lines from *Hamlet* — holds as true for podcasting as it has for every other intensely personal creative activity; let it serve as a mantra that fires up your drive to produce a terrific podcast. Honest passion, and honesty in general, cannot be faked in a podcast because many podcasters are producing their audio content without any other compensation save for ratings on various directories and analytics through Google, some feedback from listeners, and perhaps a Patreon donation. On the larger scale, podcasters, regardless of their agendas or goals, are on the podcasting scene because they want to be there, and if that honesty in wanting to deliver new content is even remotely artificial, listeners will lose interest in your podcast.

REMEMBER

Podcasting is a commitment of time and resources — to yourself and to your listeners. It's a promise you make to bring to the podosphere the best content you can produce — and with the right support, passion, and drive, your podcast will evolve, mature, and move ahead with the same zeal that inspired the first episode. Accomplishing this feat rests on remaining honest in your desire to sit behind the microphone and produce your next show. If you're not sure about the answers to "Do I know for sure what I'm getting into?" or "Do I really want to do this?" ask yourself, "If I don't want to be here, then why would listeners want to hear my latest episode?"

It's okay to ask yourself, before hitting Record, "Do I know for sure what I'm getting into?" — or, more to the point, "Do I really want to do this?" — but never ask yourself whether you *can* do this. That isn't even a question to consider. Podcasting welcomes voices of all backgrounds, all professions, all experience levels (professional, semiprofessional, or amateur). You can do this — and you have as much right as any of us. All you need is a mic, an application, a feed, and a server host. From there, it all rests on you. And once you start podcasting, produce your podcast because you want to, not because you have to.

IN THIS CHAPTER

» Discovering the obstacles of video podcasting

» Introducing streaming

» Finding out what you need to make a stream

» Creating a workflow from streaming to podcasting

» Working with HitFilm Express to edit video

Chapter **17**

One Giant Leap for Podcasting: Streaming Content

Mutation: It is the key to our evolution. It has enabled us to evolve from a single-celled organism into the dominant species on the planet. This process is slow, and normally taking thousands and thousands of years. But every few hundred millennia, evolution leaps forward.

— PROFESSOR CHARLES XAVIER, *X-MEN* (2000)

Even when you read the preceding quote, you can hear Sir Patrick Stewart's voice. Or is that just us?

We open with that prologue to the original *X-Men* film because the same thing can be said about podcasting and what we are talking about in this chapter: *streaming*. To understand how streaming impacts podcasting, it helps to look at how podcasting first dabbled with video and even "live" podcasting in the beginning.

In the Beginning: The Early Days of Video Podcasting and Live Broadcasts

In its first year, podcasting established itself as an exciting audio option. It was truly "The People's Radio" for your computer, promising a fresh alternative to cookie-cutter corporate radio. A few innovative podcasters took a closer look at the way RSS 2.0 worked and thought, "Wait a second — if we can do this with audio, what about *video?*" These brave few decided to put moving pictures to their podcasts; and while people to this day still search for just the right term (vidcast, vodcast, vodblog, vlogging, and so on), *video podcasting* came to be.

Light, cameras, and say what: The unexpected demands of video podcasting

While video podcasting presented incredible possibility for filmmakers of all levels, those new to the medium discovered that video production was not easy. Some of the downfalls to working with video are:

>> **A learning curve:** Despite what the commercials for laptops of all makes insinuate, editing video is not and never has been a push-button technology. A lot goes into video, even on the most basic of levels. With current video podcasts like *The Joe Rogan Experience* (https://podcasts.joerogan.net), *Technorama* (https://www.chuckchat.com/technorama), *Happy Hour from the Tower* (http://www.happyhourfromthetower.com), and *Universe Today* (https://www.universetoday.com), or digging deep into the archives with original video podcasts like *Tiki Bar TV* or *Ask a Ninja,* the bar to run with the big boys and girls of video podcasting is pretty high.

>> **More production time:** With audio, the aim is to create the cleanest, best polished auditory experience. When you work in video, as seen in Figure 17-1, you have a lot more to contend with than just unwanted noise pollution. Now you wrestle with lighting, camera angles, wardrobe, make-up, and so on. Editing also proves more difficult because you can't just cut a flub out without the host seemingly jumping around. Podcasting audio can prove daunting in its production needs, but video demands a great deal of time, attention, and resources.

>> **Episode frequency:** The added demands of production also makes posting of new content a real challenge. Even a biweekly posting schedule poses an incredible undertaking with casting, pre-production, filming, editing, and post-production all while maintaining a day job that is probably financing your ambitious video podcast. And if your cast and crew are strictly volunteer, even a monthly posting schedule looks daunting.

FIGURE 17-1:
Unlike in audio production where you are working against unwanted noise, video production incorporates additional factors, such as lighting, wardrobe, location, and camera angles.

>> **Audio or nothing:** Some mobile apps recognize video episodes but are unable to play them back, and some platforms will not approve your feed if you feature video episodes. Spotify stipulates, at the time of writing this edition, that it syndicates only audio feeds. So while video is tempting in so many ways, you may find yourself creating content that goes unseen.

Say that you decide to take on the task of video podcasting. You've created video projects on your computer before and found out "I . . . LIKE IT!" Contending with the preceding obstacles, you may edit your video with Apple iMovie, Apple Final Cut Pro, HitFilm Express, or Adobe Premiere to create your next podcast.

And once the episode is ready and done, you'll face a whole new set of challenges.

>> **File size:** A 1-minute video clip shot at a high definition resolution (1920 x 1080 pixels) comes out to roughly 50MB, give or take a few megabytes. 50MB for one minute of HD video. And you want to create a 15-20 minute video episode. "How do I get my file size smaller?" podcasters asked on various discussion boards about video. Although the reply may have come across as snarky, it was the only available answer: "Make a shorter episode." That old idiom of "You can't fit 25 pounds of sugar into a 5-pound bag . . ." applies aptly to video podcasting.

OH, YOU TEASE!

Sure, video can take a lot longer to record, edit, upload, and so on than an audio-only counterpart. We don't deny that part. The great thing is that there are ways to make quick, short videos that you can use to promote your podcast whether audio or video! Many podcasters have turned to services like Twitter Live and Facebook Live to make short vignettes, or *teasers,* to promote a show with larger production. While at your favorite convention, expo, or weekend event, you may find yourself gathering interviews or some other footage for your regular show. Take a few minutes to tell your audience what you are working on. Grab your mobile phone and do a quick broadcast, or put together a quick 1-minute moment (like this conversation Tee is having with his studio mascot, Benedick the GamerCat, during a tense moment in *The Last of Us*) that offers a sneak peek as to what you have planned for your next creative project.

» **Bandwidth demands:** Bandwidth, especially if your podcast is 500MB per episode (and remember, that's 10 minutes of video), can prove to be costly to content creators. If you garner 1,000 listeners, for example, you've gone through 50GB of bandwidth just for that episode. It's even heftier if your podcast is a runaway success. If you have 2,000 subscribers in your second week, you have burned through nearly 1000TB of your monthly bandwidth just for your first episode. What about the weeks — and the listeners — to come? What does your webhost offer, and can it handle this kind of traffic for your podcast?

Video podcasting still happens in various stages, but most of the productions going live are usually backed by major studios or production houses. Some incredibly innovative filmmakers are creating some amazing productions, but these podcasts are few and far between.

Going live (if you're lucky): Early attempts at live podcasting

Alongside video podcasting, even more ambitious podcasters wanted to attempt live podcasting. These hosts wanted the ability to connect with their audiences and interact with them in real-time, and some ambitious vendors promised podcasters the ability to take voice calls, interact with viewers, and export everything straight to a compressed video format or MP3 audio.

That is what was promised, anyway.

Live podcasting in the early years (pre-2010), either audio or video, was incredibly unreliable and dependent on your computer's processing power. Even with cutting-edge hardware, results may not be what you would expect. Tee still recalls when asking about a vendor's software not working properly, the response their rep gave him was "You need a better Mac." This was 2007, and Tee's Mac ran a 3GHz quad-core processor with 32GB of RAM, Apple's top of the line computer at the time. But, according to the vendor, he needed a better Mac.

Yeah. Okay, Boomer.

For those who could afford it, streaming services were available, and the bill for these services were somewhere in the six-and-seven figures. Even then, watching these streams were device-dependent, meaning if you had a slow Internet connection or older computer, you would have to continuously refresh your browser and then maybe — *maybe* — the live stream would right itself. Any live events on the Internet remained reserved for production companies with deep pockets, and only those with high performance computers could watch and interact with the hosts.

Streaming Media: Podcasting Evolved

The same year Tee was told to get a better computer, Justin Kan, Emmett Shear, Michael Seibel, and Kyle Vogt created Justin.tv, a website to allow anyone to broadcast video online. Justin.tv laid the groundwork for what is known today as *streaming*, striking a balance between data compression rates, bandwidth, and ISP upload and download rates. This opened the door for anyone to stream content live and interact with viewers. User accounts were called *channels*, and channels were encouraged to broadcast a wide variety of live video content. It was in the summer of 2011, Justin.tv decided to spin off a new streaming service that focused on video games, from their General Interest platform. This spin-off was rebranded as *Twitch* (https://twitch.tv, pictured in Figure 17-2) and quickly eclipsed Justin. tv in popularity. Eventually, Justin.tv and Twitch merged; and today Twitch is returning to its roots, still known for streaming video game content but also growing in popularity are streams featuring everyday life, special events, cooking classes, and yes — *podcasting*.

FIGURE 17-2: Twitch.tv (https:// twitch.tv) is a platform known for video games like *Sea of Thieves*, but has recently begun a return to its roots, offering podcasters a platform to connect with listeners and interact with them in real time.

A less technical definition of streaming would be a platform offering consumption or broadcasting of content, either time-shifted or in real time, on a global platform. (Sounds familiar, doesn't it?) With the development of broadband and technology able to handle robust data streams, streaming services now offer both audiences and content creators broadcast-quality entertainment online. Twitch is not alone in streaming platforms. In recent years, other streaming platforms have gone live and are now popular with content creators of all backgrounds. These platforms include

>> **Periscope / Twitter Live** (https://periscope.tv) — The first mobile streaming platform to appear, *Periscope* launched in March 2015 fully integrated with Twitter. When you go live on the app, you can send a quick notification to everyone in your network that you are live and streaming. A few years later, Twitter's mobile app offered a live feature that notifies your network that you are streaming.

>> **YouTube** (https://youtube.com) — Powered by Google, YouTube was the original hub for creative video works. Everything from simple home movies to ambitious filmmakers creating all kinds of incredible, independent works (like https://bit.ly/PunisherDL and https://bit.ly/PNEshort), but with the ability to stream either directly from your smartphone or mobile device, or from a studio in your home or office, YouTube made streaming original content (including a recording of your podcast) possible.

>> **Facebook Live / Facebook Gaming** (https://facebook.com) — The success of Twitch and Periscope laid the groundwork for Facebook launching (in August 2015) *Facebook Live,* a streaming platform fully integrated with the largest social media platform in the world. Like Periscope, Facebook Live will notify your network when you go live, even offering a tiny inset video in the lower corner of your app or browser window, taking you to the stream with a single click. *Facebook Gaming,* a variation of Facebook Live, grants viewers the ability to interact with gamers similar to Twitch.

You may want to launch a streaming channel for a lot of reasons, but the simplest reason to stream, is that it's just plain fun! Tee began his own Twitch channel on September 6, 2017, and through all the breaks, the changes, and the modest numbers he's seen, he still relishes the moment he goes live, as seen in Figure 17-3.

FIGURE 17-3:
Tee takes his love of video games, writing, cooking, and podcasting to his own Twitch channel at https://twitch.tv/theteemonster.

All this is interesting, you're probably thinking, but where are we going with all this? This is a book on podcasting. And yes, this is a book about podcasting and how podcasting and streaming work hand-in-hand in creating content and how streamlining your workflow is worth your time and attention. Tee wondered what took him so long in incorporating streaming with his own podcasting routine.

So how does one stream content? More importantly, how does one record a podcast on a stream?

Good news: If you're streaming, most of the hard work is done.

Going Live . . . for Realz This Time: Streaming Your Podcast

A couple of things you need to know about what's covered in this chapter. We are making the bold assumption that you have set up your podcasting studio and are ready to record video. The steps from here are Twitch-centric, but they are easily applied to other platforms like YouTube or Facebook.

You need a few things set up and ready to go for your first stream:

» Camera (or multiple cameras) installed

» An active account on a streaming platform

» OBS or Streamlabs OBS installed

If you are thinking "Wait, what?" then we recommend you pick up a copy of Tee's *Twitch For Dummies*, which goes in to more detail on what you need to make video happen, such as cameras, streaming software, and additional computer upgrades.

The first step: Setting up a streaming account

As we are talking about *Podcasting For Dummies*, there is a possibility — especially if you are starting from the beginning — that you are not set up on a streaming account. If you are, good on ya for being ahead of the curve! Don't worry if you aren't. We're going to get you started on this adventure in streaming content by setting yourself up on Twitch. In a few minutes, you will be up and running.

1. **Go to http://twitch.tv and from the top-right corner of your browser window select the Sign Up option.**

You can still watch Twitch streams without being signed up with the platform. However, if you want to take advantage of the Chat features, you need an account.

2. **Come up with a username for yourself on Twitch.**

   Your username is how you will appear in Chat. Your username can be a nickname you go by, a play on words, or your own name. You can approach the username in a lot of different ways. Just make sure that you are not violating any Terms of Service when you create it.

3. **Create a password or a passphrase and confirm it.**

   While passwords should be difficult to crack but easy to remember, *passphrases* are now recommended by cybersecurity experts as upping the difficulty level for being hacked. More characters are involved in passphrases, and if you take something easy like "I am a podcaster" and rework it as "!am@ p0dcasteR" for your password field as seen in Figure 17-4, you have created a very strong password that meets many platforms' criteria.

4. **Enter in your birthday.**

   Again, the date you enter is based on an honor system, but the birthday is there to verify your age, which is within the Terms of Service (TOS) as established by Twitch. For more on Twitch's TOS, visit `https://www.twitch.tv/p/legal/terms-of-service/`.

FIGURE 17-4:
Setting up a Twitch channel takes only a few minutes and is free.

5. **Enter a valid email.**

   This email is where all notifications and any news from Twitch are sent.

6. **After reviewing the Terms of Service and the Privacy Policy, click on Sign Up to complete the application.**

   Congratulations! Your Twitch account is now active.

Technically, from here, you are ready to start streaming. You have a place on Twitch, but presently the state of your Twitch account is a lot like the state of a Twitter account newly launched where the profile and bio are blank. You will want to take some time to complete your profile and set up a channel with all the unique bells and whistles needed to make the best first impressions when people show up. (For details, see *Twitch For Dummies*.) With a basic account up and running, we recommend taking a tour of Twitch. Pay a visit to live streamers running your favorite video game, hosting an open chat session, or recording a podcast, to get an idea of this new platform.

Setting the stage: Working with Streamlabs OBS

In *Twitch For Dummies*, Tee talks in depth about *Open Broadcaster Software* or OBS (http://obsproject.com/), an open source application that turns your computer and any audio-video components associated with it into your own broadcast studio. In this edition of *Podcasting For Dummies*, we're focusing on *Streamlabs OBS* (http://streamlabs.com), a variation on OBS where a variety of add-ons are waiting to be implemented and a dashboard that tracks the activity of your live stream.

Incorporating a template

Along with all of the many add-ons you can implement, Streamlabs also offers up *overlay templates* that range in themes, moods, applications, and colors, as seen in Figure 17-5. With just a few clicks, your stream is transformed from the basics to breathtaking; and, if you have the processing power to back it, your overlays and widgets can be animated, adding an even more dynamic look to your channel.

WARNING

OBS and Streamlabs OBS are not the same application. While Streamlabs did develop from OBS, the latest version of OBS has different features and functionality from Streamlabs.

The best attribute of Streamlabs is how insanely easy it is to get cracking with it.

1. **Go to** `http://streamlabs.com` **and click on the Download Streamlabs button.**

2. **Launch Streamlabs OBS once it's downloaded onto your PC or Mac.**

   Shortly after you launch Streamlabs, log in with Twitch, YouTube, or some other streaming service.

3. **Log in with your platform.**

   The Streamlabs applications syncs with the website.

4. **In the top-left section of the Streamlabs OBS application, click on the Themes option.**

5. **Find a theme for your stream either by searching through the various templates offered or by using the Scene Theme Category located on the left-hand side.**

   You can click on a template's preview image to see a full-screen preview of it.

   Streamlabs offer *Scene Theme Category* filters that narrow down search parameters to templates matching your mood. If you are known for streaming adventure or FPS games or if you are hosting a talk show, filters make the decision process a little easier.

FIGURE 17-5: Streamlabs OBS Themes section offers you a variety of moods, looks, and atmospheres for your stream. All of these themes are customizable, too.

6. **Once you find a theme you like, click on the Install Overlay button at the top, right-hand side of the template's preview.**

You now have a template in place.

Adding sources in your template

After you have a template, you need to populate it with all your incoming video and audio sources in order to get your stream up and running. This is something Streamlabs makes incredibly easy by knowing where your resources are in your setup and a series of clicks.

1. **Click on the Editor option in the upper-left section of Streamlabs OBS.**

   If you have multiple templates installed in Streamlabs, you can access all of them from the drop-down menu pictured in Figure 17-6, which will show the current active template.

   The template, if it has a Starting Soon scene, begins with an introduction screen. As seen in Figure 17-7, under the name of the template, you see other items or Scenes listed. *Scenes* are the various stages of your stream. From introduction images to intermission placards, scenes should follow a progress for your Twitch channel.

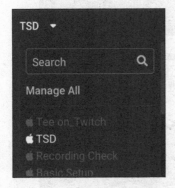

FIGURE 17-6:
The current active template.

2. **Click on the Live Scene scene (or what the template calls the scene where your stream will happen) to edit it.**

3. **Look into the Sources window and select the Background source.**

   You see the background image surround itself with a bounding box.

   Sources are exactly what they sound like: sources of audio and video needed to make your stream happen. Instead of the console managing it, you and Streamlabs are managing all sources independently of one another and using Streamlabs as something like a mixer.

FIGURE 17-7:
Scenes are different segments of your stream, and in OBS, you segue from one to the next whenever you want to go to different segments.

4. **Click the + to add a source to this Live Scene scene.**

   The Add Source window appears.

5. **Select the Video Capture Device option and click on the Add Source button.**

6. **In the Add New Source field, type Host Camera for the name of the source and click on the Add New Source button.**

   This is your Properties window to tell Streamlabs where the signal source is coming from.

7. **From the Device drop-down list, select your webcamera and click the Done button.**

   When Streamlabs detects a signal, it drops the source directly into your scene's layout. You can adjust this in the next step.

8. **Grab the bottom-right handle of the video you have made live in this scene and click and drag the video to resize it.**

9. **In the Sources window, click and drag Host Camera to the top of the list of your sources.**

10. **Repeat Steps 5 through 9 to incorporate any additional cameras and audio sources.**

    If you are using a mixer, you would have your mixer as incoming audio. Otherwise, it would be your USB microphones.

In bringing your podcast into this template, you also have an idea of how to customize your template to your own specifications. Personalize this template by editing the various labels provided. From here, you can tinker around with various elements to bring more depth to your productions. The more you dive into Streamlabs, the more possibilities there are for you. It's now a matter of what you want to accomplish and how you want to present yourself on stream.

And we're live! Podcasting in the moment

After you have your podcast set up to stream, you're ready to go live. If you are working with friends remotely or even with friends in your studio, you will want to have everyone ready to go 15 to 30 minutes before your scheduled go time in order to run final checks with audio and video. Your pre-show prep should also cover any notes you or others in your show want to talk about.

With everything agreed upon, questions answered, and equipment checked, it is now time to hit your Go Live button:

1. **Launch your recording software.**

   Part of your pre-show prep should also be making sure your DAW is recording and ready to go.

2. **Start recording on your audio recording software.**

3. **Go to your Streamlabs and confirm that you are in your Starting Soon scene; click the Go Live button in the lower-right corner of the app.**

TIP

   Instead of starting a stream straightaway at the scheduled time, streamers tend to launch a stream a few minutes before in order to give people time to receive notifications and arrive on your channel. Tee usually gives his Stream Starting Soon slate 10 minutes before fading down his introduction music and beginning his podcast.

4. **Begin podcasting, making sure to check your Chat window during the show for any comments or questions that may come up.**

5. **When you sign off on your podcast, wish your Chat audience farewell and thanks for hanging out on your live stream; then end your stream.**

6. **Stop your recording software.**

Welcome to the podcasting workflow that includes streaming. You have an audio recording of your podcast, but you also have a video stream that can be downloaded and edited either as a special video episode or stripped of its audio and edited as a backup recording.

It's a stream, it's a podcast. It's a stream and a podcast!

So what exactly does streaming the recording of your podcast bring to your production? Why exactly is going live with video essential to an audio podcast?

>> **Live interaction with your audience:** The element of a live audience adds a level of excitement and energy to your production. Something you or one of your co-hosts say can spark commentary from your audience. A special guest with a background interest (gaming developer, Fantasy author, voice actors) may inspire questions. This is the ride you take with interaction that a live audience provides by streaming.

>> **A platform where a new audience will find you:** As we mention in Chapters 12 and 13, we are always looking to expand our listener base, and you may find that with streaming. New visitors to your Chat may want to catch up with previous shows, find out how your show began, and how far you have come in your own experience with what your podcast is all about. And bringing an audience into the show is a great way to attract new listeners.

>> **Video featuring bonus content:** Once your stream is concluded, streaming platforms offer options for the exporting of streams to other media platforms like YouTube (seen in Figure 17-8) or Vimeo. You can offer these uploads as embedded videos in your show notes. The raw footage may include footage not found in the final podcast. You may want to consider this as a potential incentive for your audience to attend the live stream.

FIGURE 17-8: Streaming podcast recordings bring your audience right into the moment, and unedited episodes can be featured in show notes.

» **Automatic backup recording:** Redundancies are a beautiful thing, as some of the close scares documented in this title only legitimize. Too often, wonderful once-in-a-lifetime interviews and engaging discussions with co-hosts are completely lost when a recording fails. By incorporating streaming into your podcast's workflow, you immediately create a backup recording ready for production, if needed.

If you like, you can use the audio from your stream as your podcast file as well. To retrieve audio from your stream:

1. **Go to your streaming platform and access your directory of previous broadcasts after your stream is done.**

   On the Twitch platform, this is the Video Producer section. You can access previous broadcasts for two weeks. After that time period, Twitch erases the broadcasts from its servers.

2. **Find your podcast stream and access the option, seen in Figure 17-9, to download the episode on to your computer.**

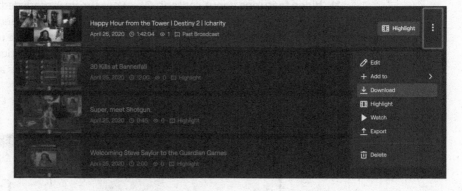

FIGURE 17-9:
Twitch offers content creators the option to download streams when they are available in your video directory.

3. **Once the episode is downloaded, import the episode into a media player or video editor you are familiar with.**

   A common media player available as a free download for both Windows and Macintosh operating systems is QuickTime.

4. **Export the podcast stream as Audio Only from the media player or video editor you are using.**

   For QuickTime, this process is as simple as opening the episode into the app and then choosing File ➪ Export ➪ Audio Only.

5. **Import the stream audio into your digital audio workstation.**

6. **Edit and save your podcast.**

SERIOUSLY? A SELFIE STICK?

The much-maligned selfie stick has been either a source of ridicule or, for those really lost in their own narcissism, a source of horrifying (if not comical) death. But when it comes to filming on your smartphone, a selfie stick saves your arm from a lot of cramping and offers your video a bit of stability. With the right angle and the right leverage, your selfie stick actually works as a collapsible steady cam and can allow you better options in capturing more people in the moment and even more activity in your shot. For additional polish to your self-shot video, you can also employ the DJI Osmo Mobile Gimbal, a steadycam solution for your smartphone. Just remember to be aware — well aware — of your surroundings. Getting the best shot for your podcast will not matter one bit if you are dead or recovering in a hospital.

And yes, Tee (on the right in the figure) was being very careful with a vintage steamtrain slowly advancing behind him.

Photo (right) credit: Babs Daniels, taken at Steampunk unLimited 2015

WARNING

When streaming your podcast, especially if you are on location, make sure you are aware of your surroundings. It is far too easy to lose your footing, bump into someone, or find yourself in a rather precarious (if not dangerous) situation. Be careful!

Streaming platforms make what was once promised back in the early days of podcasting not only a reality, but also an easy and stress-free option to video podcasting. Working with live video feels less like a leap between mediums and more of a logical and seamless addition to your workflow. Streaming a podcast may appear as if you are making a serious upgrade to your studio; but much like your early steps with podcasting, your upgrades don't have to break the bank.

And if you want to look more into how to stream content, don't worry — Tee's got it covered with *Twitch For Dummies*, a deep dive into his own amazing journey into streaming content.

6

The Part of Tens

Chapter **18**

Ten Types of Podcasts to Check Out

A h yes, narrowing down the thousands upon thousands *upon thousands* of podcasts out there to an elite ten: The Top Ten Podcasts You Should Be Listening To. The Best of the Best. The Select Few to Follow.

No pressure.

Where to begin? Where to begin? With each edition of *Podcasting For Dummies*, your humble authors sat down and put together our own top ten lists of podcasts, and here's what we have noticed: There are a *lot* of podcasts out there. While some of our listening habits have remained steady and constant as the Northern Star, our taste can sometimes shift into unexpected offerings . . . which, if you listen to *Astronomy Cast* (http://www.astronomycast.com), you would know Polaris is not *that* constant. We have found, over our years of podcasting, that we listen to a solid cross-section of *types* of podcasts, not just in the way of genres but in production values. So with a quick edit of the outline and a change of mindset, we present Ten *Types* of Podcasts to Check Out.

There are other kinds of podcasts that are, perhaps, not showcased here; but with these ten kinds of podcasts as a starting point, you can easily begin to fill your media player with a variety of offerings. When you're comfortable with what you hear, you can then follow podcasts that podcasters will recommend. Feeling even braver, you can delve into the many directories out there, some of which offer suggestions based on your listening habits.

Here's your starting point for podcasts to seek out and consume. We are confident that from the list given in this chapter, you can sample a wide cross-section of interests, passions, and projects. Relax, download, and give your eyes and ears a treat.

TIP

Always check podcasting directories to see how active these shows are before you subscribe. The podcasts in this section are constantly changing. They were actively producing content when we wrote this book, but printed material is never as up to date as a good podcast directory.

Tech Podcasts

If you're thinking about podcasting, you likely have a comfortable knowledge of computers, the Internet, and blogging. But regardless of how technologically savvy you are, like any aspect of life, you can never stop learning. That's why you should subscribe to at least one *tech podcast*.

The agenda for a tech podcast is (surprise!) technology. All geek, all the time. Geeks, nerds, wizards, and Tech Help gurus sit behind a microphone and pull back the curtain on how your computer works, how a podcast's RSS feed can be better, and how to make your time behind the keyboard more efficient.

Tech podcasts come delivered to you in a variety of skill levels and on a variety of topics:

» **Macs:** Some excellent podcasts for Mac users include *The Mac Observer's Apple Context Machine* (shown in Figure 18-1 and available at https://www.macobserver.com/show/apple-context-machine) and the *iMore Show* (http://www.imore.com/podcasts), covering the latest developments coming from Cupertino, latest product reviews, and the odd opinion or two on the direction of Apple.

» **PCs:** *Windows Insider Program* (http://windowsinsider.mpsn.libsynpro.com) is an ambitious effort spearheaded by Windows Insider chief Dona Sarkar to podcast updates from within the Windows Insider community, whereas the *MSPoweruser Podcast* (https://mspoweruser.com/podcast) is more end-user/consumer focused to help you be more productive on your PC (and all things Microsoft).

>> **General computing:** Many of the tech podcasts out there like *This Week in Tech* (or TWiT, as it is so lovingly referred to by its hosts and listeners, available at http://www.twit.tv/twit) or *Daily Tech News Show* (http://www.dailytechnewsshow.com), are generic in their approach to operating systems, following a mindset that a computer is a computer and the rest is mostly bells and whistles.

>> **Technology perspectives:** *Computer Talk Radio* (http://computertalkradio.com) and *Technorama* (http://www.chuckchat.com/technorama) give personal and (in many cases) light-hearted perspectives on technology in society and go beyond the geek-speak, addressing issues on how technology can affect just about everything.

Whichever podcast you feed into your podcatching client, find a tech podcast that's right for you and either enjoy the new perspective or allow yourself to grow into your geekdom. There is a lot to learn about computers, smartphones, the Internet of Things, and other cool Q Branch gizmos that are available at your local electronics store. When you have the basics down, these podcasts allow you to unlock their potential and go beyond your expectations.

FIGURE 18-1:
The Mac Observer's Apple Context Machine is one of many podcasts offered to Mac users to get the most out of their iPads or MacBooks.

Independent Media Podcasts

Podcasting is audio and video content on demand, and after taking in what is being offered on the radio, that is a very good thing. Maybe it's a hazard of getting older, but some of us just aren't hearing anything on the air that's all that interesting or exciting. Unless you are really into auto-tune or really hungry for yet another hacker-crime-investigation-unit-comprised-of-incredibly-pretty-people series.

The good news is that media mavericks are alive and well and doing just fine in the 21st century, finding a new promotion channel with podcasting.

Independent labels, where the artists also work as promoters, producers, and holders-of-all-rights, have control over where their music plays, how often it is played, and how much it will cost you. Many indie musicians who are having trouble getting exposure and radio airplay will grant permission for podcasters to use their music. What does this cost the podcaster? A few moments of time and a spot or two, such as, "This music is brought to you by . . ." and "Visit this band online at w-w-w-dot . . ." And in return, the musician is exposed to a worldwide audience.

Musicians such as The Gentle Readers, Michelle Malone (both featured on *Evil Genius Chronicles* at `http://www.evilgeniuschronicles.org`), George Hrab, and Rubber Band Banjo (both featured on Tee's Parsec-winning podcast *Billibub Baddings and The Case of the Singing Sword*) have all enjoyed the benefits of associating themselves with a podcast. Catching wind of this, indie artists of all backgrounds are turning to podcasts in order to spotlight their independent works whose vision might go against the corporate entertainment industry's notion of "what the public wants." Here are a few such podcasts you may want to check out:

» **Welcome to Night Vale** (`www.welcometonightvale.com`): Best described as *A Prairie Home Companion* meets *The Twilight Zone* and known for its quirky, sometimes sinister storytelling, the popular podcast has proven itself a fantastic platform for independent label music. Whenever Cecil announces the Night Vale weather report, listeners are treated to music from artists such as Anais Mitchell, Robin Aigner, daKAH Hip Hop Orchestra, Destroyer, Daniel Knox, Toys and Tiny Instruments, and Eliza Rickman, just to name a few.

» **Irish and Celtic Music Podcast** (`http://celticmusicpodcast.com`): Celebrated musician Marc Gunn hosts *The Irish and Celtic Music Podcast*. It's all in the title. Marc offers his podcast as a spotlight for the best in Irish and Celtic culture and sound; and with over 20,000 downloads per episode, multiple wins for "Best Podsafe Music" from the People's Choice Podcast Awards, and a consistent spot in iTunes' "What's Hot" directory for music podcasts, The Celtfather shows no signs of stopping.

>> ***Making Movies is HARD!!!*** (http://spindryproductions.com/podcast generator): The independent film industry — particularly with the popularity of video podcasting, YouTube, and other online video providers — is embracing podcasting platforms as not only a place to promote but a place to educate. *Making Movies is HARD!!!*, hosted by Timothy Plain and Alrik Bursell, discusses independent filmmaking, but not just shooting and editing a film. Timothy and Alrik deep dive into all aspects of filmmaking. Writing. Producing. Directing. Even how you take care of your cast and crew. Timothy and Alrik also cover the downsides of making movies: hours of investment, rejection, lost opportunities, self-doubt. It's all here.

>> ***Making Lemonade with Jordan Morpeth*** (http://www.jordanmorpethart. com): He creates comic books. He loves talking about *Star Wars*. He hosts a podcast about life and how it all impacts his pursuits. This is Jordan Morpeth, illustrator and graphic designer in Sydney, Australia, and *Making Lemonade* is his audio journal that sometimes explores the challenges in art and design, the in's and out's of the comic book industry, or simply the magic moments happening all around him. If you are a creative individual and want a perspective on your passions, enjoy a glass of lemonade with Jordan.

Science Podcasts

The approach to science made popular by Bill Nye the Science Guy, Mr. Wizard, and the Mythbusters, is rampant in the podosphere. Here are some science podcasts that we recommend:

>> ***Astronomy Cast*** (http://astronomycast.com): If you've always wondered about what's really going on beyond our atmosphere, take a listen to a podcast descendant of Carl Sagan's *Cosmos* concept. For those who would never leave the planet without their attitude, Fraser Cain and Dr. Pamela Gay dubbed the show *Astronomy Cast*.

>> ***Science Magazine*** (http://sciencemag.com): When it comes to science, you don't get much closer to the source than to *Science Magazine*, the official podcast of the award-winning *Science* journal. This podcast is presented by the staff and contributing voices of *Science*, offering news and policy from the world of scientific research.

» **_The Brain Science Podcast_** (http://brainsciencepodcast.com): When it comes to the human brain and its mechanics, Dr. Ginger Campbell, MD, knows her way around it. _The Brain Science Podcast_ (pictured in Figure 18-2) features the latest books about neuroscience as well as interviews with leading scientists from around the world. Neuroscience, you might think, would appeal to a niche audience; but with over 7.5 million downloads, Dr. Campbell's podcast continues to educate the world about how our brains work.

FIGURE 18-2:
On _The Brain Science Podcast_, Dr. Ginger Campbell, MD, delves into the most mysterious and incredible of computers ever created: the human brain.

» **_Therapy for Black Girls_** (http://www.therapyforblackgirls.com): Hosted by Dr. Joy Harden Bradford, _Therapy for Black Girls_ focuses on making mental health topics more relevant and accessible for Black women. Dr. Bradford taps into pop culture to illustrate psychological concepts in order to overcome the stigma surrounding mental health issues. This podcast aims to make mental health topics accessible and relevant.

Science shows are good for you — like vegetables, only better. Not only can you learn something new or broaden your scope in a field you may regard as a hobby, but you can marvel at how the hosts demystify concepts once comprehensible only to PhDs and keep the ideas easy to grasp.

Self-Development Podcasts

Why not broaden your horizons — personal or professional — with podcasting? For example, say you've always wanted to learn Spanish. You could shell out some bucks for the "Teach Yourself Spanish" series, or you could listen to the free *SpanishPod101* podcast (`https://www.spanishpod101.com`). *SpanishPod101* is a podcast that is part of a multiple device approach across PDFs, online forums, and smart devices that culminate into a survival guide for a new language. With its podcast as the first step, *SpanishPod101* helps you navigate situations at the restaurant, at the bank, or if your car breaks down.

If you aren't interested in learning another language, take a look at some of the other subjects offered in the podosphere. Here are just a few examples of what's available:

>> **Education:** Podcasts can be like your favorite class after the bell has rung, offering material completely devoted to what you might not have heard while in middle and high school. *Stuff You Missed in History Class* (`http://www.missedinhistory.com`), hosted by Tracy V. Wilson and Holly Frey, offers details behind Lady Jane Grey, Executive Order 9066, or comedians Abbott & Costello, each episode an in-depth look at events from our past. Instead of history, what if you are working on your language skills using the *Duolingo* (`http://www.duolingo.com`) app? If you are enjoying the lessons (and if you're on the app, you know it's a blast learning a language this way!), you can increase your immersion with the *Duolingo Podcast*, specific to what language you are learning.

>> **Career Development:** Are you looking to transition to working from home, but wonder if you're missing a step somewhere in staying productive? Have a listen to *WFH The New Reality* (`https://anchor.fm/wfhpodcast/`), hosted by Ian Scott, a podcast about making the right steps forward in a career that involves a commute from your bed to a Macbook. For female entrepreneurs, the *She Did It Her Way* podcast (`http://shediditherwaypodcast.com`) features interviews with risk-takers who worked hard to bring their ideas to reality. And when it comes to incredible visionaries — Elon Musk, Norman Lear, and Michelle Obama, to name a few — the *TED Talks* podcasts (`http://www.ted.com/talks`) offer up all kinds of great career advice from those who shape the world.

>> **Personal Development:** *TED Talks* don't just stop at Career Development (as there are many podcasts TED produces), but also help you better yourself and your life. Are you looking to change your eating habits or maybe just find calm within the stress of daily life? Podcasting can bring a healthy

alternative to your current digital lifestyle, and maybe even lower blood pressure and cholesterol points in the process. To find a bit of health and harmony in your life, check out the *Mind of Snaps Podcast* (`http://mindofsnaps.com/podcast`) hosted by Jessy (also known as SheSnaps, pictured in Figure 18-3), a gamer, blogger, and streamer, dedicated to encouragement, balance, and overall positivity. For those wanting their health and philosophy peppered with a bit of Bruce Lee, dive deeper into a martial art you study (or are thinking about studying) with *whistlekick Martial Arts Radio* (`http://www.whistlekickmartialartsradio.com`).

FIGURE 18-3:
Mind of Snaps helps you manage life and work stress through meditation, organization, and overall positivity.

>> **Money management:** If you're looking for advice on managing your money, check out *Money Girl* (`http://moneygirl.quickanddirtytips.com`). Or if you are trying to manage finances after you have made the ultimate life investment (Hint: It involves the words "I do" or a variant of that), try *His & Her Money Podcast* (`http://www.hisandhermoney.com/podcasts`), which focuses on joint financial plans and their balance in a strong, marital relationship.

There is a lot to learn from the world, and whether you're tuning in to one of numerous podcasts sponsored by universities and colleges or to an enthusiast who wants to share and swap resources with you, all this continuing education is available online, in audio, and at no charge.

MANAGER TOOLS, TAKING A PODCAST FURTHER

This section wouldn't be complete without mentioning *Manager Tools* (https://manager-tools.com) and its spinoff, *Career Tools*. Whether you are a manager or individual contributor, you're sure to find something of value in each episode. Working from home, resumes, communications, they've got it all. *Manager Tools,* the company, began a weekly podcast in June 2005 targeted at giving managers specific, actionable information. Shortly thereafter, *Career Tools* was launched as a spinoff for individual contributors with the same goal. Both shows became very popular and are free. One of your authors has been a loyal listener since the beginning.

However, the podcasts themselves are not the end game. The *Manager Tools* teams use them as lead generation to their courses, workbooks, and conferences, which pay the bills. Additionally, you can purchase a personal license to access the complete archive, written show notes for each episode, a roadmap on how to roll out their tools, and their online interview tool (which alone is totally worth the price).

Comedy Podcasts

There are times in life when you just need a good laugh. A real bonus with comedy podcasts is having those good laughs categorized, digitized, and waiting for you, only a tap or click away.

Humor, though, is in the eye of the beholder and the ear of the listener. Performing a search on "Comedy Podcasts" will offer you many, many choices, ranging from kid-safe comedy to performing arts to adults-only discussions with an irreverent approach. Comedy covers a lot of ground, but straight-up comedy podcasts are not intended to do anything other than entertain.

The way these podcasts make you laugh covers a wide spectrum of how humor is defined. Here are a few comedy podcasts that we enjoy:

>> *Comedy4Cast* (www.comedy4cast.com): The unpredictability from *Comedy4Cast*'s host, Clinton Alvord, is one of the elements that adds charm to the show. Sometimes you get a tongue-in-cheek review of an electronic gadget, another time you hear a parody of another popular podcast, or Clinton could be offering a comedy serial. Whatever is on the menu provided by cast and crew, *Comedy4Cast* has but one goal: having you pull over to the side of the road, so you don't drive into the ditch, or setting the hot coffee aside so as not to spit it on your keyboard. DISCLAIMER: *Comedy4Cast* is not responsible for damages incurred while listening.

>> *The FuMP* (http://www.thefump.com): *The FuMP* stands for "The Funny Music Project" and features a cavalcade of comedy musicians that include Devo Spice, the great Luke Ski, Worm Quartet, Power Salad, Nuclear Bubble Wrap, Carrie Dahlby, Carla Ulbrich, Robert Lund and Spaff, Raymond and Scum, and more! All of the music showcased in *The FuMP* is protected under a Creative Commons license so listeners know they can share the music or even make their own music video to it, so long as no one makes money at it. *The FuMP* also operates as a network, offering website visitors the links to artists for additional laughs beyond the podcast.

>> *Alison Rosen is Your New Best Friend* (http://www.alisonrosen.com/ariynbf): This podcast began as an online talk show, streamed from journalist Alison Rosen's Brooklyn apartment. This window into a comic writer's life evolved into interviews that center around people's genuine struggles, both in her backyard and around the world. *Alison Rosen Is Your New Best Friend* has been described as a marriage between *Seinfeld* and *Charlie Rose*, but Rosen considers her podcast as a chronicle of daily life and how we tend to laugh at what we as humans deal with.

>> *The Geologic Podcast* (http://www.geologicpodcast.com): No, wait, *The Geologic Podcast* is a podcast featuring the music of musician and author

George Hrab! Hold on — *The Geologic Podcast* is a scientific and skeptic podcast featuring the musings of musician and author George Hrab! Actually . . . yes, *The Geologic Podcast* is all this and a comedy podcast guaranteed to get you thinking all while rocking your socks off. There is a lot happening in this podcast, so best hold on to something when you dive in.

Slice-of-Life Podcasts

Comedy is prevalent in all the various genres of podcasting, especially with the podcasts that just take a look at life and give it a perspective. Sometimes you need humor to deal with loved ones, life's unexpected pitfalls, or just the world on a whole. Slice-of-life podcasts offers audiences a chance to laugh at the headlines, slow down with an in-depth look at a specific topic, or offer a look at a lifestyle. The subjects covered are varied, but the podcast serves as windows into the hosts' world. True, most podcasts are; but in these podcasts, it's about the moment, the here and now, and what is happening in their lives. From this casual and candid approach comes a podcast that might make you laugh, might make you cry, and maybe — just maybe — make you think.

» **Truth Be Told** (https://www.kqed.org/podcasts/truthbetold/): Hosted by Tonya Mosley, *Truth Be Told* offers itself as that friend who is always there to celebrate your victories, listen to your gripes, and offer you a shoulder to cry on. This podcast gives a candid look into the multifaceted lifestyle of being black in America.

» **The Way I Heard It with Mike Rowe** (http://mikerowe.com/podcast/): Mike Rowe, best known for his snarky-but-sincere approach to America's working force in *Dirty Jobs*, brings all his dry wit and endless charm to *The Way I Heard It*. This podcast tells new and slightly unexpected stories of celebrities, history, and icons of culture, passed on from person to person, generation to generation. It's a quick listen — usually under ten minutes per episode — and is guaranteed to make you smile.

» **The Innovative Mindset Podcast** (https://izoldat.com/): Produced by author, speaker, and entertainer Izolda Trakhtenberg, this podcast is a weekly talk show with business leaders, entrepreneurs, and creatives on tapping into innovation and problem-solving strategies. This show may sound better suited for our Self-Development Podcasts section, but Izolda's *Innovative Mindset Podcast*, pictured in Figure 18-4, is more about how the show's guests discover solutions from every day inspirations. *The Innovative Mindset Podcast* emphasizes the importance of remaining human in a digital world, and finding a fulfillment in life through *"Eureka!"* moments big and small.

WARNING

As it goes with slice-of-life podcasts, the general content can go in any way: G, PG-13, to "What did they just say?!" Always check the podcast's listings (in directories or on their websites) for an Explicit tag to see how "playfully blunt" these podcasts can get.

Gaming Podcasts

You might think that podcasts about games and gaming would be something better suited for the Slice-of-Life, Geek, or Comedy sections, depending on how lively these style of podcasts get. Long-form and serialized productions of friends gathering together to either crawl through a dungeon, claim the title of *King of*

Tokyo, or share strategies and latest news from *World of Warcraft* or *Minecraft* stand in a class all their own. Gaming Podcasts are quite popular, and the skill levels featured on these podcasts range from the professionals appearing on the circuit to the novices purely in it for the fun.

>> *Esportsmanlike Conduct* (https://www.esportsmanlikeconduct.com): You may not believe that eSports — professional teams with such names as Cloud9, Evil Geniuses, and Digital Chaos that come together to play video games — is a thing, but it is. It is a *serious* thing; and *Esportsmanlike Conduct*, hosted by Brandon "Atrioc" Ewing and Nate Stanz, features interviews with outstanding athletes from the digital sports arena, latest news and results from recent tournaments, and breakdowns of how teams performed.

>> *How Did This Get Played?* (https://www.earwolf.com/show/how-did-this-get-played/): Some really incredible video games are out there — *God of War, Destiny, Maneater, The Last of Us, BioShock* — but did you ever wonder about those games that make you go "How in the name of Ralph Baer did this game ever get into development?" From bad mechanics to ridiculous premises to "Did I just do that?!" Heather Anne Campbell and Nick Wiger put their deep love for video games to the test with *How Did This Get Played?* featuring the worst, the weirdest, and the FML of console and PC gaming.

>> *The Steamrollers Adventure Podcast* (http://riggstories.com/the-podcast): Michael J. Rigg loves to write. Michael J. Rigg loves role playing games. Michael J. Rigg also loves steampunk. When Mike wrote his debut novel, *Clockwork Looking Glass*, he wanted to delve even deeper into his world, but invite his friends along for the ride. This inspired Mike to design a d20-style role playing game around his steampunk universe. *The Steamrollers Adventure Podcast* (shown in Figure 18-5), provides an immersive RPG experience for its audience as Mike provides music and sound effects associated with events in-game. There is also a lot of inappropriate humor peppered throughout which has been known to happen during RPG sessions, but "Storycrafter Mike" keeps the action going as he pulls his friends and special guests deeper into a Wonderland of his own making.

>> *Critical Role* (https://www.critrole.com): This podcast began as many role playing campaigns did: A bunch of nerdy friends playing Dungeons & Dragons in each other's living rooms. *Critical Role* is now the must listen to RPG podcast, attracting over a half million viewers every week. With dreamy Matthew Mercer as your dungeon master in their second campaign storyline, popular voiceover actors and special guests take on impossible challenges while enjoying a few laughs along the way.

FIGURE 18-5:
The Steamrollers Adventure Podcast features an original roleplaying game set in an incredible alternate past of clockwork, steam, and intrigue.

Podcasts of the Pen

As you know, Tee stepped into podcasting with an idea no author had yet set out to do: serialize a novel in podcasts. When the podcast of *MOREVI* ended for Tee in the summer of 2005, he looked at the audio equipment he had and asked himself, "So . . . what do I do now?" By this time, there were a few writerly podcasts out there all talking about the *craft* of writing, but there were no active podcasts on the *business* of writing books. To fill this void, Tee launched *The Survival Guide to Writing Fantasy* in September 2005, a podcast about the everyday operations of a writing career.

The last episode of *The Survival Guide to Writing Fantasy* aired in 2009, but many more podcasts have appeared that cover all aspects of writing from the business to the basics behind getting the work done to marketing and promotion that works.

>> **The Creative Penn Podcast** (http://www.thecreativepenn.com): *New York Times* and *USA Today* bestselling thriller author Joanna Penn has developed quite a reputation with *The Creative Penn*, a podcast (and companion blog) that goes into all aspects of writing. Business. Creative demands. Time management. If you have questions about the writing industry, Joanna is a one-stop

shop of knowledge. Her podcast also features interviews with authors of all backgrounds and all genres, so if you are looking for help in your writing career or creative project, *The Creative Penn* may be your first stop.

>> *Writing Excuses Podcast* (http://www.writingexcuses.com): A Hugo-winning podcast featuring seven professional authors and a huge collection of interviews, advice, and back-and-forth banter, *Writing Excuses* is a fast-paced, weekly podcast for writers, by writers. They like to say, "Fifteen minutes long, because you're in a hurry, and we're not that smart. . ." but if you get them going, the show may run a little longer. *Writing Excuses* wants to help their audience become better writers, regardless if the end goal is writing professionally or writing for the soul.

>> *The Shared Desk* (http://www.theshareddesk.com): Tee Morris and Pip Ballantine love writing together, and you get the idea they enjoy podcasting together when you hear them on *The Shared Desk*. When Tee and Pip turn the mics on, their topics of discussion cover collaborative projects of all kinds, what's happening in the publishing industry, and also go into what is happening with their individual works as well. Whenever possible, Tee and Pip bring guests in studio to pick up different perspectives on writing.

>> *The Everyday Novelist* (http://everydaynovelist.com): J. Daniel Sawyer has penned over 25 books and over 30 short stories. While working on his next work-in-progress, Dan hosts *The Everyday Novelist*, a daily writing podcast about taking your writing passion to the next level. Within a few minutes, *The Everyday Novelist* tackles a new topic either presented by Dan, a special guest, or from his listeners. It's a podcast that serves as your daily affirmation for what you want to accomplish as a writer.

Geek Podcasts

If you have ever attended a Comic Con or a science fiction–fantasy–horror convention, you may have witnessed firsthand passionate individuals who love to talk about their favorite movie, television show, or game. Or maybe you caught a live interview with an actress or writer that you love, and heard that Nathan Fillion is a comic book nerd or that Erin Grey is big into Tai Chi. Truly one of the joys of podcasting is capturing good laughs between friends, luminaries, and a room full of fellow geeks attending a con and riffing on the mics.

This is the joy you hear in these podcasts. A love and a passion for all things geek. (There it is again! The *P* word — passion.)

SOMETIMES, OPPORTUNITY PRESENTS ITSELF. TAKE IT!

This is Tee Morris and Jack Mangan interviewing *Battlestar Galactica*'s Richard Hatch in February 2006 at Farpoint, a science fiction convention in Baltimore, Maryland. You'll notice that Richard is wearing a coat. He wasn't cold. He had a plane to catch, but he made time before leaving to sit down and talk on the mic about his experience between the BSG of the 1970s and of the 2000s. With *Battlestar Galactica* being a hot property once again, it was a real treat for Tee and Jack, two fans of both renditions of the epic space opera, for Richard to give them time, but how did they do it? How did two podcasters — at a time when podcasting was still a new media — land an interview with one of *BSG*'s regular players?

Simple: Tee asked.

We go into more detail on approaching people about interviews in Chapter 6, but there are those opportunities that present themselves in the moment. There is nothing wrong with asking celebrities, subject matter experts, and special guests relevant to your podcast *if they have a few minutes to talk*. There is a strong possibility that a favorite writer, actor, performer, or comic book artist might have time to sit down, but you will never know if you don't ask. The worst thing a potential guest on your podcast could say to you is "No." (And if they do, they're not being jerks. They have somewhere to be or a pocket of time to get something to eat.) But if they say "yes" then be ready to go. And as we say in Chapter 6, be ready with solid, intelligent questions.

Then let the magic happen.

» **Geek Therapy Radio** (https://www.geektherapyradio.com): Proud nerd Johnny Hemberger hosts *Geek Therapy Radio*, a podcast celebrating pop culture and many exhibits of geekdom in an open, fun, and heartfelt way. Whether it is the latest in technology or an interview with actors, writers, and film directors, *Geek Therapy Radio* is a sit-down with guests happy to let their geek flag fly high and proud.

» **Geek Radio Daily** (http://www.geekradiodaily.com): *Geek Radio Daily* featuring The Wonderful Billy Flynn, Podcasting's Rich Sigfrit, and The Flynstress provides a daily (yes, *daily*) dose of geek every weekday. **Mondays** highlight the weekend box office results, **Tuesdays** cover what's out on Blu-Ray, **Wednesdays** is New Comic Book Day, **Thursdays** are all about new video game releases, and **Fridays** offer up what's opening at theatres everywhere. On alternate weekends, as a bonus, Carol the Cat hosts a review of geeky Horror Movies as part of her *Scream Queens Survival Manual*. Then there is the *GRD's Weekly* super-sized show — in many cases, recorded live at a convention — featuring special guests and in-depth discussions on how to get your geek on.

» **Wired Magazine's Geek's Guide to the Galaxy** (https://www.geeksguide show.com): *Geek's Guide to the Galaxy*, hosted by author David Barr Kirtley and produced by *Lightspeed Magazine* editor John Joseph Adams, is an unabashed embracing of your inner geek. *Geek's Guide* features in-depth conversations and interviews over books, movies, games, and comics, all coming from a place of science-fiction, fantasy, horror, and a love for the genre.

Podcasts about . . . Podcasting

It may sound redundant and feel a little odd to listen to a podcast about podcasting. It would be like having an electrician come over to another electrician's house to fix faulty electric wiring. But why not?

» **The Feed** (http://thefeed.libsyn.com): When it comes to podcast tutors, you want to work with people who have trustworthy track records (and thank you for trusting us!); and when it comes to trustworthy track records, Elsie Escobar, Rob Walch, and LibSyn are responsible for many podcasts available today. *The Feed* (shown in Figure 18-6), is LibSyn's official podcast, covering podcast strategy, tips, media hosting, and all things LibSyn (See Chapter 3 for more). As *The Feed* is a community-driven show, their topics go beyond getting a podcast off the ground. Rob and Elsie go deep into ways of growing your audience, promotion that won't come across spammy, and crafting a strategy for your podcast, whether it is a casual endeavor or a professional venture.

FIGURE 18-6:
The Feed, hosted by Rob Walch and Elsie Escobar, not only keeps you in the know on changes coming to LibSyn, but also helps you produce better podcasts.

>> ***The Audacity to Podcast*** (http://theaudacitytopodcast.com)**:** Podcaster Daniel J. Lewis is your host for *The Audacity to Podcast,* an award-winning podcast all about podcasting using Audacity and WordPress. Along with those foundations, Dan recommends audio gear, suggests good podcast habits, and reminds you of the reasons why you podcast in the first place.

>> ***For the Love of Podcast*** (https://fortheloveofpodcast.com)**:** Welcome to *For the Love of Podcast,* a production from Billy Samoa Saleebey, an award-winning filmmaker and host of the interview podcast *Insight Out. For the Love of Podcast* offers beginners and long-time producers solutions and strategies that help showrunners up their game. With over 1,000,000 podcasts now live across various directories, Billy features voices that will help you keep pace with the platform, reach new audiences, and keep your passion to podcast alive and thriving.

Chapter **19**

Top Ten Reasons to Podcast

"Podcasting is just a fad."

"I've listened to a few podcasts, and I'm not impressed."

"This is nothing more than streaming media, and that really hasn't gone anywhere."

The naysayers of podcasting (especially those who have no idea what it is, have never listened to one, but are the first to dismiss it) have a plethora of reasons why podcasting will fail or never *really* catch on in mainstream media.

And yet here we are, over a decade later and on the fourth edition of *Podcasting For Dummies,* and we still have naysayers talking smack about podcasting, even though podcasting is more popular than ever.

We podcast for a great many reasons. You have heard us talk about those reasons, but this chapter — this final word from your authors — is our hard sell. We love doing this. We returned to *Podcasting For Dummies* because we enjoy podcasting that much. Heck, Tee launched a brand new podcast as we were writing the third edition in 2017, Chuck started a new *Star Trek* podcast called *The Topic Is Trek* (http://thetopicistrek.com) during final edits and has another planned for his day job before this edition hits the shelves, and Chuck and Tee are resurrecting another podcast after finishing this edition!

So, yeah — there is something to podcasting, and there are some good reasons to jump into the podosphere with us.

Here are a few (ten, to be exact) reasons why.

You Are Considered a Subject Matter Expert

Guy Kawasaki is a name you should know in social media circles. As if his time with Apple Computers (where his team was responsible for the marketing of the Macintosh back in 1984), his *New York Times* bestselling titles (like *The Art of Social Media, The Art of the Start 2.0, Reality Check* and *Rules for the Revolutionaries)*, and how his blogs remain in the Top 100 visited blogs in the world wasn't enough to make him a name in tech circles, Kawasaki has cemented himself as something of an oracle when it comes to influencing, both online and in the walking world. Some of his advice? Rock solid. Other bits of it? Worth questioning. However, in one of his most loved/hated of blogposts, "Looking for Mr. Goodtweet" (https://guykawasaki.com/looking-for-m-1), Kawasaki had this to say:

"Establish yourself as a subject expert. One thing is for sure about Twitter: There are some people interested in every subject and every side of every subject. By establishing yourself as a subject expert, you will make yourself interesting to some subset of people."

Yes, Kawasaki is talking about Twitter but this rule does apply to podcasting quite aptly. When you launch a podcast, you are establishing yourself as a voice in the subject matter of your podcasting. When it comes to writing, Joanna Penn on *The Creative Penn* brings her experience as a *New York Times* and *USA Today* bestseller along with her own experiences promoting her works and her brand to *The Creative Penn Podcast. SpyCast* (https://www.spymuseum.org/multimedia/spycast), the official podcast of the International Spy Museum, was started by and hosted for years by ISM's Executive Directory and 36-year spy veteran Peter Earnest. When Peter decided to step down, he wanted to ensure the expertise remained so he turned the mic over to Dr. Vince Houghton, historian and curator at the museum, his specialty being in intelligence, diplomatic, and military history, with expertise in late World War II and the early Cold War. Pictured in Figure 19-1, *SpyCast* continues to be the authoritative podcast on intelligence gathering.

When you launch a podcast, you speak with the voice of authority. You speak as an expert in your field, as someone who has a proven track record and an individual who knows a thing or two about the topic of discussion. Speak with confidence. You have a lot to say, and what you have to say makes a lot of sense.

You Are Passionate about the Subject

You may not be the world's greatest gamer. You may not be the best at crafting costumes. You may not be the world's fastest runner. But if you are passionate about a sport, if you are passionate about creative endeavors, if you are passionate about a board, card, or console game, then yes, you should be podcasting about it.

It is a reoccurring theme in this book, and it bears repeating as many people want to podcast about something they love but are intimidated by the amount of work that *could* go into a podcast. Another obstacle is for passionate people who podcast to compare themselves to more polished, professional podcasts that gather the best and the brightest guests in-studio. How do you compete with productions like that?

Well, Chuck and Tee have seen professional podcasts come and go, sometimes after eight episodes. Sometimes, after only two. This occurrence is known as *pod-fading.* Why do productions podfade so quickly? In many instances, podcasts are regarded as revenue generators. In other words, these hosts are in it for the money. Podcasting can be a money-making venture (as discussed in Chapter 14) but that "overnight success" rarely happens. Eventually, after repeated attempts at producing that magic viral episode, the studio lights are turned off and the equipment is packed away.

But while there are plenty of podcasts out there for Bungie's video game, *Destiny*, it's no secret that Tee, Nick, and Brandon love the game. It was their passion that led to *Happy Hour from the Tower*. With so many podcasts out there about technology, why do Chuck and Kreg continue to podcast *Technorama* since 2005? Because after hundreds and hundreds of episodes, *Technorama* continues to nurture that passion. No show ever sounds forced or trite. There's a genuine joy within every podcast.

You can be an expert in your chosen field, or you can just be a huge fan. Passion should be at the core of every podcast. Without that, you can't really find the drive to sit yourself behind the microphone and record, only then to edit and produce the final work for your audience. So if you feel the drive to podcast, do so. It will take you far.

You've Got a Creative Itch to Scratch

Maybe you never thought of yourself as a creative person, or maybe you were a creative person when you were younger. Maybe there's been an inspiration working at the back of your mind, and you've been wanting to explore it. The weird thing about this idea, this unexpected muse that has grabbed hold, is that you may need to pick up some skills that you don't know.

Podcasting, as we have shown, is not only something you can pick up quickly, but it is an affordable venture.

Science fiction-fantasy author Aly Grauer and game connoisseur Drew Mierzejewski took a few brave steps into podcasting with *Dreams to Become* (http://dreamstobecome.com), Aly's website and home to many of their limited series podcasts. Their own podcasting journey began with *The Disney Odyssey* where Aly, Drew, and special guests joined them in their personal journey through every animated feature film from Walt Disney. Then came *The Night's Rewatch*, a step back to the beginning of HBO's *Game of Thrones*. But still this wasn't enough so *DND20 Public Radio*, a sketch comedy created at the intersection of NPR, *Dungeons & Dragons*, and *Waiting for Guffman* launched, all under the *DTB* feed.

It was that drive and creativity that brought Aly and Drew to the One Shot Podcast Network (http://oneshotpodcast.com/), where they were tapped to launch *Skyjacks: Courier's Call* (https://skyjacks-couriers-call.simplecast.com), an all-ages actual play podcast spun off from One Shot's *Skyjacks* RPG. *Courier's Call*, pictured in Figure 19-2, which follows the three young apprentices in the Swiftwell Courier Service (played by Aly Grauer, Paulomi Pratap, and Aaron Catano-Saez) undertaking adventures in the skies above Spéir. Alongside their own professional pursuits in entertainment, *Skyjacks: Courier's Call* serves as another outlet for Aly and Drew, with the podcast as their stage and their imaginations allowed to thrive; the end result is an unforgettable production.

Once you have your studio, either a simple audio setup or something more complex, a podcast serves as your blank canvas for whatever creative endeavor you're about to embark on. This could be a throwback to the days of radio theatre or this could be a personal journey for you accomplishing a life goal like physical fitness, a college degree, or a trip across the country. This podcast is where you share your creation with the world. Regardless of whether your feedback is positive or negative, this is your stage. Assure your audience what they can expect from your feed, and then allow your creativity to run. This is your creative corner of the Internet. Make the most of it.

FIGURE 19-2: Creative power couple Aly Grauer and Drew Mierzejewski bring their creative energies to One Shot Podcast's young adult adventure, *Skyjacks: Courier's Call*.

You Like Playing with Tech Toys

Let's be honest: The toys a podcaster gets to play with are just so cool.

Microphones. Mixer boards. Gadgets for going portable. Software. The tools of the trade, while sometimes coming with steep price tags, are absolutely tempting. Not only do some of these technical gadgets stimulate the creative juices within your brain, but they can also be quite the showstoppers with the company you keep.

Tee has a terrific story about when he sat down in 2007 to interview Peter Earnest, executive director of the International Spy Museum. At this time, Peter was the host of the aforementioned *SpyCast*. Tee was using a Zoom H4 for the interview which is the predecessor to the H4n mentioned in Chapter 4. While both models differ in features, the H4 was of a similar physical design. Both models resemble a Taser, and this fascinated Mr. Earnest. Before the H4 went hot, Peter asked Tee many, many questions about the device, its features, and the quality of its recordings.

Think about that for just a moment: The curator of a museum that features a lipstick pistol, an Enigma machine, the Model F-21 buttonhole camera, a rectal toolkit (yes, you read that right), and the "Bulgarian Umbrella" used to fire a tiny pellet filled with poison, was completely captivated by this portable audio recorder of Tee's. Considering Peter's background, that is saying something about the allure of content creation gear.

If there is something to the latest technology that makes you happy and gets your blood pumping, whether it is the MXL Overcast bundle or an all-in-one recording device like the Zoom P4 (pictured in Figure 19-3) that offers you recording options, consider all the wonderful toys you find in podcasting. While this may sound like a frivolous reason to think about launching your own show, consider how your skill set also broadens. Working with gear like condenser and dynamic microphones, portable digital recorders, and software that produces audio productions will only serve to your advantage when called on to create something special for an office demonstration or for a special event at home. It might surprise you, as well, what kind of skills you pick up in producing a podcast in the ways of planning, project management, presentation skills, resource budgeting, and time management.

And to think it all starts with the tech toys.

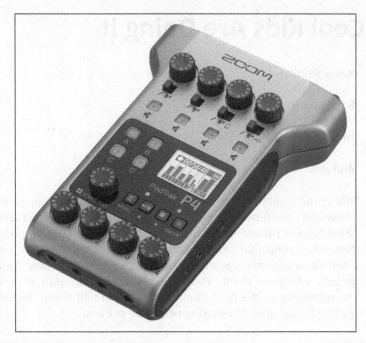

Bring Your Friends Together

You are putting together your notes for your podcast. It could be your first podcast. It could be a new podcast to add to your portfolio of podcasts. Whatever the case, you decide that instead of your voice being the only voice on the show, you reach out to a few friends in the area or online whom you know are just as passionate about the subject you plan to podcast.

Maybe this podcast is a RPG session, or perhaps you want some fellow L.A. Kings fans to get around the mics and talk about the last game. Or you invite some friends who all share an interest — writing, period costume productions, movie soundtracks — to come on over and riff about it on pod. With schedules agreed upon and set, you settle in with your friends either in real time or over Discord, Google, or your favorite online conference network of choice and record. Maybe you don't realize it, but your recording sessions are more than just your chance to herd content and build up a buffer for your show. The podcast is your guaranteed connection with you and your friends. Recording or streaming a podcast is locked-in time when good friends know they are getting together to have a little fun and share some quality time around microphones.

Another great thing about this podcast with your friends is that the podcast becomes a journal — a testament — of your friendship. That's worth the time, especially when you do retrospective episodes.

All the Cool Kids Are Doing It

Kevin Smith.

Katie Couric.

Neil deGrasse Tyson.

And you.

It's a little humbling how many high-profile journalists, celebrities, and industry influencers are turning to what was once a platform for indie artists exclusively. What is most satisfying is, after a decade and some change, podcasting is still a fantastic platform for independent creatives. For the NPR, AMC, and ESPN types, the podcast also serves as a fantastic opportunity to go beyond their time on stage, screen, or sports event. Podcasting is something akin to a great equalizer as, regardless of the production values, we are all doing the same thing here: getting on the mics and sharing what's on our minds.

This is some great company to be in, so why don't you go on and get your podcast up and running? It's okay. There's plenty of room in the podosphere for what you've got. Bring your best, put your heart into it, and get your pod on!

I Can Do More

In the 2009 edition of *Star Trek*, Captain Christopher Pike says to young upstart James T. Kirk:

"Your father was the captain of a starship for 12 minutes. He saved 800 lives, including your mother's and yours. I dare you to do better."

Pike's words serve as a great mantra for podcasters as with every show produced, podcasters look to do better. We look to improve. We look to grow. Some podcasters, after running a show for a time, love to look back on early shows and see the progress made from those first steps. It is said amongst some podcasters that the first five episodes of any podcast (even those done by experienced hosts) will suck, but they are allowed to suck.

Chuck and Tee do not necessarily adhere to or believe in that rule as some podcasts find their voices straightaway within two or three episodes. Others, we have found, record an "Episode 0" just to see if the idea looking awesome on paper translates as well to media. Then you have shows that find a groove once the mics are hot or the camera goes live. Whenever you find your voice, creating something that people react to is truly special and worth the time and effort. The slippery slope in podcasting, though, is that drive and desire to do more.

Take a look at Figure 19-4 – Tee's years in podcasting since 2005. He's the first author to podcast a novel from cover to cover. He took what started as a marketing strategy and turned it into a book. Then he goes on and creates a podcast about the business side of writing, launches companion podcasts for his books in social media, and helps establish a website where other podcasters can podcast their own novels and collections.

FIGURE 19-4:
Tee Morris, podcasting from 2005 to today.

And yet, he wanted to do more.

Tales from the Archives is launched, opening up the steampunk universe he created with his wife, to other authors and to wider audiences. *The Shared Desk* is soon launched after this, offering commentary on the latest news in the publishing industry as well as a behind the scenes look at an author's life. While delving into the current edition of this book, Tee follows an impulse and launches *Happy Hour from the Tower*, a loving nod to the Bungie game *Destiny*. All this, and he still hosts alongside Chuck *Podcasting For Dummies: The Companion Podcast* in order to keep this book fresh and up-to-date.

This may seem like a lot for Tee to take on, but for Tee this is a real love for the platform and the medium. He challenges himself to do better. He desires to do more.

This is the drive behind a podcaster.

Bring Out the Best in You

When you sit down to create a podcast, you want the production to rival that of professional broadcasting. No, you don't have that budget but that doesn't mean you don't strive for that level of polish and professionalism. Even with a show like *Technorama* (seen in Figure 19-5) which comes across as spontaneous and off-the-cuff (and for the most part, it is!), Chuck and Kreg work to make their show a production that podcasters and radio show hosts all strive to reach, if not surpass. There is a sense of accomplishment and achievement in producing a podcast that gets people to stop and ask, "Wait, hold on — you do this in your home?"

FIGURE 19-5:
Chuck Tomasi (left) and Kreg Steppe (right) with special guest, Dr. Pamela Gay (center), at a live recording of *Technorama* at Dragon Con, a show that always brings out the best in its hosts and the many guests they entertain.

Sometimes, we do. Sometimes, we take our act on the road. It all depends on where we can kick up the most trouble!

Podcasting encourages by its artistic and technical nature to compel producers and show hosts to create the best podcast in whatever subject the producers and hosts pursue. Does that mean it will be regarded by audiences worldwide as the best? It depends on how you measure your success. Most podcasters measure the impact of a podcast on ratings and rankings. Others care more about feedback from their listeners. Some measure their podcast's success by how long their show runs after the premiere episode drops. This also compels producers and hosts to insist on creating the best podcast they can. This is how podcasting engenders a real desire in those involved to offer a show that is as much fun for audiences to listen to as it is for the podcast's crew to record, edit, and release.

Let podcasting bring out the best in you.

Talk to Interesting People

Bugs Bunny was right:

"My, I bet you monsters lead iiiinnnnnteresting lives. I said to my girlfriend just the other day 'Gee, I bet monsters are iiiinnnnnteresting,' I said. The places you must go and the things you must see — mmmyyyyy staaaaarrrs. And I bet you meet a lot of iiiinnnnnteresting people, too. I'm always interested in meeting iiiinnnnnteresting people.
— BUGS BUNNY, "HARE RAISING HARE" (1946)

Now while Bugs was referring to monsters, Chuck and Tee refer to really *iiiinnnnn-teresting* people. Some of the people the authors of this book have met over the course of podcasting include authors who have made an impact in their genre (Robert J. Sawyer and Terry Brooks); actors who have plenty of stories from behind the scenes (Richard Hatch and Lani Tupu); other podcasters who have made lasting impressions in and beyond the podcasting community (Grant Baciocco and Dr. Pamela Gay); and even scientists who have changed the world in what we know about it (Dr. Robert Ballard discovered the *Titanic* wreckage in 1985). Both Tee and Chuck consider themselves fortunate for meeting a wide range of guests in their years of podcasting.

Sometimes, though, you are lucky enough to talk to guests who wind up becoming far more familiar than just guests on your podcasts. A perfect example is Chuck and Tee. In the infancy of podcasting, Tee reached out to Chuck, asking to be on *Technorama* as a guest. Tee returned for other promotional opportunities, at first, but those return trips led to meetups at conventions, which led to friendship, which eventually led to over a decade of *Podcasting For Dummies* editions like the one you are reading now.

Not all the people you meet will lead to lifelong friendships, but through podcasting you will meet a lot of *iiiinnnnnteresting* people who will in some way impact your life. Some of those impacts you will notice straightaway. There will be those discussions you have with people who become a more subtle touch on your life, and you might not notice it until years down the road. If you're really lucky, you could be Patrick G. Holyfield who, after being taken away from the podcasting community by cancer, is remembered fondly at the P.G. Holyfield Meat & Greet, pictured in Figure 19-6. At this event, podcasters old and new come together to do what Patrick enjoyed most: Create friendships.

Each person you meet, though, is part of a network, and that network — personal or professional — will at the very least broaden your view of the world. Your experiences in podcasting may catch you completely off-guard and will enrich your life for years to come.

FIGURE 19-6:
The P.G. Holyfield
Meat & Greet, a
yearly meetup
hosted at
Balticon, brought
podcasters
together to
remember this
fallen podcaster,
a testament to
the lives he
touched.

The Ultimate Thrill Ride

There is something scary, humbling, and intimidating about taking something you have created and releasing it to the world. You think it's good. Good enough to share, even. But once you release your podcast out into the world, it is out there. For everyone to consume. And for everyone to critique, criticize, and dissect. This is a whole new level of fear when your first show goes live.

It is also an amazing rush of adrenaline, euphoria, and accomplishment.

Tee has been podcasting for over a decade. This is nothing new to him, and yet he can attest on launching *Happy Hour from the Tower* that he was terrified beyond reason. Why? This was a whole new kind of podcast for Tee. He had never podcasted about video games before. This was his first regular show with multiple cohosts. And when it came to a subject matter — Bungie's award-winning video game, *Destiny* — Tee was not the best of players, let alone "well known" in the game's community. Oh, and as the show's launch date was less than a month out from Bungie revealing details of *Destiny 2*, it just seemed a bit late to launch a new podcast about a game that had been on the market since 2014.

Still, Tee launched the podcast, and he's been enjoying the ride since that first show dropped.

Podcasting, whether on the grand scale or a small, personal stage, is an adventure. The longer you podcast, the bolder you become. The bolder you become, the more you want to test your limits. You find yourself reaching out to experts in the subject matter of your podcast, or maybe you reach out to the hosts of podcasts similar to your own. You invite others to appear on your show. Or you find yourself carrying recording equipment everywhere, much like a photographer does with camera gear. You set up your portable studio, fire up the mic, and begin documenting. You talk. You meet new people. People with stories to tell. Suddenly, you find your own personal network growing. Your network grows closer, and those contacts become friends. Then, if you are lucky, those friends become family.

Podcasting is an incredible ride, and it can take you to unexpected places. It is rewarding, even the podcasts that never seem to take off. They are an education of what to do differently and how to improve. And just when you think you've got it all figured out, new technology and new approaches appear, and you find yourself at Square One all over again. Possibilities are endless, and the unexpected — regardless of how much you plan — will happen. This only adds to the fun ahead.

Now it's your turn. You've got an idea. The microphones are waiting. Go on and hit Record.

And hold on. Your adventure is officially underway.

Chapter **20**

Ten Original Podcasters

I f you're reading this book, you're likely new to podcasting and might have missed out on the trials and tribulations that went along with that first huge growth spurt between 2004 and 2005. It was truly undiscovered country not only for Chuck and Tee, but also for the people listed in this chapter, recognized as some of the original voices who have stood the test of time. While some of these "O.P.s" ("Original Podcasters") may not still have their original shows, they all have been podcasting since the "early pioneer days" and continue to produce content.

As you read these names, try to remember that these are regular people like the rest of us. True, some of them may seem larger than life, but all of them put their pants on one leg at a time — unless they're wearing kilts.

Although we encourage you to make your own decisions, a future podcast star would do well to take a listen to the wisdom/information/rants put forth by this group. Often controversial and usually informative, each of these people has a unique outlook on the world of podcasting. We've found these folks to be helpful guides in our own podcasting careers. But (as we've said probably way too many times in this book for our editors' tastes) your mileage may vary.

Mignon Fogarty

Mignon, better known as "Grammar Girl" to her listeners, is a master; and one of many reasons Chuck and Tee asked her to kick off the previous edition of *Podcasting For Dummies*. She has taken the topic of English grammar, something most would find mundane — and many think they have mastered only to find out they learn something new in every episode — and turned it in to a podcast empire.

Trivia time! What was Mignon Fogarty's first podcast? If you said *Grammar Girl*, then you have chosen poorly. A little-known fact, she had a podcast in 2005 called *Absolute Science* cohosted with Adam Lowe. That show ran for about 8 months and while it didn't prove to be quite the success its successor would become, it did teach her some valuable lessons.

Her breakout show, *Grammar Girl's Quick and Dirty Tips for Better Writing* (http://www.quickanddirtytips.com), started in 2006 and received numerous acclaims and recognition such as #2 on iTunes's most recommended podcasts list. As she describes it, "It will help you improve your writing and delight you with the quirks of language." Her grammar show has since turned into a *Quick and Dirty Tips* network featuring topics such as productivity by the "Get-It-Done Guy," Steve Robbins; money and finance by the "Money Girl," Laura Adams; and even parenting by the "Mighty Mommy," Cheryl Butler. Mignon even got her old cohost, Adam Lowe, from *Absolute Science* back to be the original "Modern Manners Guy." She's been featured in the *Wall Street Journal* and on *The Oprah Winfrey Show* and *The Today Show*. Twice.

Like many podcasters, Mignon's audience is not contained to the United States — take that, terrestrial radio! She receives feedback from all over the globe, again not all that unusual, but always appreciated. What's interesting is that many of the comments from her listeners are from teachers using her episodes to teach English as a second, third, or fourth, language.

Her podcast gave birth to the book *Grammar Girl's Quick and Dirty Tips for Better Writing* (Holt) in 2008 and reached number 9 on the *New York Times* Bestseller List. And during the writing of this edition, her title *Grammar Girl Presents the Ultimate Writing Guide for Students* was Macmillan's No. 1 bestselling nonfiction book at Barnes & Noble, and the Grammar Girl podcast celebrated 14 years of podcasting with close to 800 episodes produced.

When asked why she continues to podcast the unsurprising answer is, "I really love it. I get to learn new things along with the listeners, and it's always great when I hear how I helped someone. That happens almost every week. What could be better than that?" Mignon Fogarty is one of the lucky few who get to earn a living by podcasting.

Adam Curry

In 1987, your authors were finishing school, and MTV (the icon of the generation that remembers when they actually played music videos) hired a new VJ named Adam Curry. Decades later, despite life experiences, diplomas, and other strange changes and artistic ventures, it's still baffling to think that the same guy who was introducing the latest Madonna video is now referred to as "the Podfather."

Adam Curry's importance to the craft might be the single best example that illustrates the "just-do-it" nature of podcasting. It was Adam Curry, not some highly trained programmer, who hacked together the first podcatching client program to harness the power of the recently created enclosed-media files in RSS 2.0 feeds — and made them downloadable during the computer's off hours.

Since that time, Adam has become the most recognized (and arguably the most popular) podcaster around. While his first podcast, *Daily Source Code*, has since podfaded, Curry continues to produce *No Agenda* (http://www.noagendashow. com), cohosted with John C. Dvorak. *No Agenda* listeners never know quite what to expect: a tour of Adam's kitchen, a discussion over the latest round of talks in a business venture, or sharing a content partner that will change the shape of podcasting. Everything and anything is possible.

It's worth making time to listen to *No Agenda*, even if you fast-forward through the parts that don't appeal to you.

Mur Lafferty

Podcasting has never been a "Man's World," as female podcasters were blazing trails in 2004 alongside the boys. Some of those leading ladies of the RSS feeds have either podfaded or moved on to other pursuits, but one remains to this day as an outstanding podcast personality: Mur Lafferty (http://murverse.com) or "The Grand Dame of Podcasting," as she is referred to with great affection by this book's authors.

Mur Lafferty's podcasting career began with *Geek Fu Action Grip*, a show that was an ongoing commentary about what was going on in her life, what games she was playing, and what bands she was listening to. Each episode of *Geek Fu* ended with an original essay of something that struck her funny bone. This podcast spawned *I Should Be Writing*, the 2007 Parsec-winning podcast documenting her own tests and trials in getting her first novel published. As proof of her talent, she has been a finalist or a Parsec Award winner in 2008, 2010, and 2011. She was a finalist for a Hugo Award in 2012 and won a Hugo award in the category of John W. Campbell award for Best New Writer of 2013.

She has never stopped raising the bar for podcasters. Keeping many of her podcasts to the basics, Mur produces quality content from both sides of the brain: the analytical (*I Should Be Writing*) and the creative (The *Heaven* series, *Playing for Keeps*). Mur, with her talent in writing, also raises the bar for other creative podcasters, inspiring many to podcast their own fiction and even return to a writing passion they had stepped away from for many years.

It's hard to listen to podcasts and not hear the name of Mur Lafferty. It's harder still to subscribe to any Mur Lafferty podcast and not be completely taken by her. Her personality shines in every episode, with every work, and with every show.

Steve Boyett

Like many featured in this chapter, Steve Boyett's background was not in radio, broadcasting, audio engineering, or even computers. Instead, Steve's early interests were in writing, where he boasts several books, short stories, and novellas to his name. Go look him up at http://www.steveboy.com! Several years ago — okay, maybe more than several — he started composing electronic music for fun and began DJing at parties, which he refers to as "straight-up crack."

Almost like the proverbial accidental invention, Steve stumbled on to his fate. "Because I'm a writer, I was considering a fiction or interview podcast. Doing a music podcast hadn't occurred to me." On a complete whim, he asked his girlfriend, an aerobics instructor and runner, if she thought anyone would be interested in some of the workout mixes he was producing. Her response was something like this: "Are you kidding?! Do you know how much I used to pay for this as an aerobics teacher? It's perfect!" So he launched *Podrunner* (http://podrunner.com), a regular, hour-long dose of fixed-tempo music for walkers, joggers, and runners, from beginner to competitor. There were plenty of music podcasts at the time, but Steve was the first to do a music series specifically for workouts. At first he thought he might attract 50 people or so.

Talk about underestimating your audience!

iTunes featured his podcast within days of it being listed, and his website crashed within days of *Podrunner*'s debut. It became so popular that it shot up and stayed in the Top 10 for about a year. That's when he gave some serious thought to legitimize it or kill it. Thankfully he chose to invest and continue.

At its height, *Podrunner* was commanding over 600,000 downloads a month, receiving sponsorship from Timex and the United States Navy. The podcast still produces new and original mixes, but for Steve, the benefits do not stop at the success of his BPS-bending compositions. "I have met people that I wouldn't have otherwise met. A listener reached out to me and asked me if I'd be interested in DJing for a group of fire dancers at the annual Burning Man Festival where they had sound-activated flame cannons, a 30-foot stage, and on and on. I said, 'You had me at sound-activated flame cannons.'"

But the biggest reward he receives to this day is from hearing how he helped people. Almost every week he gets a note about someone who had lost 100 pounds thanks to his music, or a chemotherapy patient who said, "This got me out of my house and on the way to recovery." And since the first edition of *Podcasting For Dummies*, Tee still takes *Podrunner* with him on 5K runs, 10K runs, and (at the time of this writing) the odd half-marathon. Steve continues to make the Finish Line appear that much sooner for Tee.

When asked why he continues to podcast after all this time, at first Steve jokingly said, "I'm in complete denial about its influence on my life," but later admitted, "It's what I do. The thought of giving it up really bothers me, and when I do think about it, I think of those listeners whom I've helped."

Adam Christianson

All superheroes have their origin story. Adam Christianson, the Clark Kent of Podcasting, is really the Superman who brings us *Maccast* (http://maccast.com). His story starts with a degree in Printing. (That's right, kids, the paper-and-ink kind of printing that came before the Internet.) Adam has always nurtured an interest in Apple Macintosh computers dating way back to their introduction, and it just so happens that's what his printing company was using. As time went by he found himself being the de facto IT guy until he approached the president of the company and made it official. Over the years, his Mac-strength continued to grow. In the mid-90s, he discovered HTML and found himself interested in web development while learning all he could about Macs.

In late 2004, he was listening to some of the early shows and thought, "I need to find a Mac podcast to help pass the time on my hour-long commute (in San Diego)." Sadly, he found out that there were no Mac podcasts at the time.

(Insert screaming sound effects here.)

Inspired by *The Dawn & Drew Show* (http://thedawnanddrewshow.com), Adam recognized how easy podcasting could be on discovering his favorite podcast was powered by a Logitech USB microphone and GarageBand. Adam thought, "I've got GarageBand — I'm halfway there!" On December 13, 2004, the *Maccast* was revealed to the world. As he states at the beginning of each episode: This was ". . . the show for Mac geeks, by Mac geeks."

That last part, *by Mac geeks,* is important to note because his original intent was to create a 5- to 10-minute daily news show about all things Mac. Remember, this is before the days of the iPhone and iPad. Macintosh computers and early iPods were about all Apple had going at the time! What amazed Adam was that after his very first episode, a listener emailed him with a technical question. This may not seem too crazy today, but in those days he wondered, "How did this guy even find my show?!" Back then, podcast directories as we know them today didn't exist. Finding new shows was generally done by word of mouth — other podcasters or listeners mentioning you. It was as if he sent a message in a bottle and got a response back from across the ocean.

Since that time Adam has expanded the stories he covers to include a wider range of Apple products, but he still keeps the name and tag line the same — mostly by listener request. The audience has also grown considerably, allowing Adam to be his own boss splitting time between podcasting and web development — with the bulk of revenue coming from the *Maccast*.

After talking to Adam, one quickly gets a sense of his humility. He's not concerned with being perceived as the authority on Macs or having millions of listeners. His main motivation for doing the show is the sense of community and the feedback it generates. He considers the *Maccast* as a "global user group" where he does just as much learning as educating.

Keep up the great work, Mac of Steel!

Dave Slusher

Dave Slusher is one of those lucky individuals who happened to be at the right place at the right time. As he describes it, "I started podcasting partly out of convenience and dumb luck." In 2004, he had been following the latest advancements of transferring and listening to audio files on the Internet, but it wasn't until Adam Curry's first podcatching client written in AppleScript that things really started to heat up for him. "It was only a matter of time before someone put it all together. The pieces were all there and sooner or later it was going to happen," says Slusher. Once that happened, Dave got out the gear from his former radio gig and started the *Evil Genius Chronicles* (http://evilgeniuschronicles.org) in late August 2004 — one day after Adam Curry started his *Daily Source Code* — and became what is now one of the three longest continually running podcasts.

When asked to describe his show, Dave's response is, "It's about music, technology, and culture — but I know that doesn't sound too compelling." Maybe that description is a bit lackluster because underneath the podcast, Dave's perspective is what makes *Evil Genius Chronicles* work. He conveys situations that affect him, digs for the deeper meaning, and articulates it in a way that matters to the listener. Another interesting aspect about *EGC* is that you never really know what you're going to get in each episode. Like many things, Dave seems to fly in the face of convention and make the most of it. Remember that note about not shaking up your show format too much or you'll confuse your listeners? Well, let's just say Dave has a four-letter response to that too, but it may not work for everyone. One episode may be a combination of music and monologue, the next could be an interview with a comic book artist, and another could be a panel discussion with friends. It's how he rolls and how his podcast thrives.

When asked why he continues to podcast, he said, "I can do whatever I want unfettered. My show, my format, my content. I'll keep doing it while it's still fun. Besides, as long as I continue, I still have the status of one of the longest running podcasts. When I stop, I'll just become a footnote to history — and nobody wants that."

Scott Sigler

When it comes to the early days of podcasting fiction, Tee has been described as epic classical music like Wagner. Mark Jeffrey could be compared to jazz music of Dave Brubeck's style.

Then you have Scott Sigler (http://scottsigler.com). He's Metallica. *The Black Album*.

"I podcast because it's the perfect vehicle for serialized fiction," Sigler says of his decision to podcast his debut novel, *Earthcore*. He seeks not only to hook listeners with the format, but also to keep them in suspense until next week's episode: "My novels are long, with cliffhangers at the end of each chapter. That's designed to keep you turning the pages. With podcasts, you just have to wait for the next episode."

Scott's nostalgia for radio-style serials is not the only motivation behind his podcasts. "The amazing reaction I get from the listeners is the motivation," he adds, acknowledging his loyal fan base, also known as *Junkies,* as they became hooked on his podcasts. The fans showed Scott (along with print publishers Dragon Moon Press and Crown Publishing) exactly how hooked they were when the podcast of *Ancestor* premiered in print. The book shot up Amazon charts to hit #1 in both Horror and Science Fiction. That was enough for Crown to approach Scott with a contract for another one of his podcasts, *Infection,* which the publisher retitled as *Infected.*

For Scott, the passion had a purpose; but even on achieving the coveted *New York Times* Bestseller List, Scott continues to deliver high octane fiction crossing many different genres, one podcast at a time, as part of Empty Set Entertainment. "To know that my fiction has entertained thousands of people, helped them escape from their day-to-day lives, and just plain have some fun means the world to me. Is it a power trip? Sure! But knowing that I've delivered a great story and entertained someone is a fantastic feeling."

Michael Butler

Original — that's the word that comes to mind when talking to Michael Butler, host of the *Rock and Roll Geek Show* (http://www.americanheartbreak.com/ rnrgeekwp/). Before meeting Michael you might have visions of a hard rockin', loud mouthed, swaggering, crazy man, only to find out he's a very humble down-to-earth guy. He states, "I don't listen to any other music podcasts because I don't want to be influenced by them. I'd rather just do my own thing."

Michael earns his place in this chapter because he is somewhere in that first ten or so podcasters who immediately followed Adam Curry. He started in early September 2004 and hasn't stopped since — okay, if you are reading this book in the year 2220, he stopped. At the time, he was playing in a band called American Heartbreak and running a blog for them. He came across Adam Curry's audio blog and was inspired by what Adam was doing. Like many podcasters just dipping their toes in, he started with modest technology. He was talking into the built-in speaker on the laptop. Shortly afterwards he upgraded to a USB microphone and eventually got a mixer and a bit better recording equipment.

Michael describes his show as "A guy talking about music, playing music, reviewing albums, and doing interviews." Michael is very well known in the podosphere not because of the popularity of his show, but because of his humbleness and quick wit. In fact, contrary to his on-air personality, he's a very quiet and shy person, often stating when he's at social events, "I'm the quiet guy in the corner."

Get to know Michael Butler and you discover his resume includes working full time for Adam Curry and collaborations with many other podcasters, including Dave Slusher on the show *Mad at Dad* (http://madatdadpodcast.com).

Despite the popularity of *The Rock and Roll Geek Show*, Michael really doesn't care about the numbers. In fact, he states, "My fans are probably some of the most loyal I have ever encountered." Hardly an episode goes by where he doesn't do a segment called "Opening Butler's Mail" to find anything from cash to special hunting arrows that Ted Nugent uses. That's some serious fandom! He continues to podcast largely because of the fans and "It's just fun to do."

Rock on, Mr. Michael Butler! Rock on!

Dr. Pamela Gay

Let's just go on record here — science is cool. At least these two authors think so. That makes Dr. Pamela Gay, the wit and wisdom behind this edition's Foreword, one of our favorite podcast veterans. Like Mignon Fogarty, Dr. Gay brought academics to the podcasting world with that right blend of education and entertainment. Also like *Grammar Girl*, Dr. Gay's hugely successful show *Astronomy Cast* (http://astronomycast.com) was not her first dip in the pod-pool. The birth of Pamela's podcast universe started when she was on the way to a meeting of the American Astronomical Society in early 2005 when her friend Aaron Price said he was reading about this new thing called podcasting. He told her about some religious evangelicals doing a show to convert people, to which she replied, "We can convert people to science!" Within a week they had the first episode of *Slacker Astronomy* online. The show ran for about 18 months until life intervened, taking the show's hosts in different directions. Not one to let her podcasting passion sit idle, she quickly connected with colleague Fraser Cain and started *Astronomy Cast*, "A weekly facts-based journey through the cosmos where we help you understand not only what we know, but how we know what we know." Great tag line, by the way!

Astronomy Cast, and its cousin podcast, *365 Days of Astronomy*, have been finalists or winners of Parsec Awards from 2007 to 2012, and now Dr. Gay can be seen on occasion in episodes of *The Universe* explaining something like how the universe ends. She has also been seen as a speaker at science and sci-fi conventions. Both on her podcast and in real life, Dr. Gay does not shy away from being a fan as well. "I was talking to Richard Hatch, the actor who played 'Apollo' on the 1970s series,

Battlestar Gallactica, about the meteor that struck the Earth's atmosphere over Chelyabinsk, Russia in 2013," she recalls. "I was happy to share with him his intersection between science and science fiction. The meteor's origins were from an Apollo asteroid."

Geek points to you, Pamela!

And if you wonder if Dr. Gay knows what she's talking about, the "Dr." in her title is her PhD in Astronomy from the University of Texas, so yes, she knows what she's talking about. What's even better is she makes it fun and interesting to listen to. This is another topic that can come across (to some) as dry and boring, but once again the hosts' passion for the topic makes it a treat to consume.

Brian Ibbott

We put Brian Ibbott on this list for two reasons. First, he's one of the original music podcasters, and you might be thinking of starting a music podcast of your own. (Hey, lots of us wanted to be DJs when we were kids.) "It's weird hearing my name associated with the word 'original,'" jokes Ibbott. More importantly, Brian is a music podcaster who is doing it right — the legal way.

As mentioned in Chapter 5, podcasting licensed music legally is a challenge. With his podcast *Coverville* (http://coverville.com), Brian works directly with ASCAP, BMI, with the musicians, and other organizations that hold the rights to major-label music and songs. Although many music podcasters are just a process server away from a major lawsuit, Brian is sitting pretty.

Brian also serves as an example of how to create a niche podcast out of the music you love and fill a void not covered by the traditional outlets. Brian has a passion for cover songs and has built a show completely around songs that fit that bill. "I wanted to create a radio show that I always wanted to listen to," he states. He took this concept a step further by featuring groups of cover songs sharing a common thread, thereby producing themed episodes of his podcast. He's even started his own record label! Simply brilliant — we don't know of any radio stations or programs that provide this service.

Brian also has a charitable heart. Each year on the last Friday in November (known as Black Friday to many) he does a 24-hour live streaming marathon of cover songs, known simply as *Cover-thon*, to raise money for Alzheimer's research.

And *Coverville* is still producing episodes to this day. What a legend, this guy!

So before you rush out to make the next hot music podcast, consider the legal ramifications. And while you're pondering that, also ask whether your future podcast sounds like the same stuff people can listen to over their radios. Follow Brian's lead and give the world something different with your podcast.

Authors' Footnote

As we wrote this chapter, we recognized that there were far more podcasters from the early days out there still producing original content. This was, arguably, a difficult list to compile and chapter to write; and this final chapter took us weeks to accomplish. We regret that talented O.P.s did not find themselves in this chapter and acknowledge their hard work, persistence, and content that continues to find subscriptions in podcast client apps everywhere.

Index

Numbers

About the Authors

Tee Morris: Tee is a communications professional based out of the Washington, D.C./Virginia/Maryland metro area and an award-winning author of science-fiction and fantasy. Tee's 2002 historical epic fantasy, *MOREVI: The Chronicles of Rafe & Askana,* was the first novel to be podcast from cover-to-cover and led to his team-up with podcaster Evo Terra on the original *Podcasting For Dummies* in 2005. Since then, Tee has written other social media titles, including *Twitch For Dummies, Discord For Dummies,* and *Social Media for Writers,* penned with his wife, Philippa Ballantine. Both Tee and Pip continue to podcast together with their Parsec-nominated podcast *The Shared Desk* and continue to write together in *The Ministry of Peculiar Occurrences,* a steampunk series that has won several Parsec awards as well as numerous literary awards, including RT Reviewer's Choice for Best Steampunk of 2014. Alongside Nick and Brandon Kelly, Tee also hosts *Happy Hour from the Tower,* a podcast celebrating their mutual love for Bungie's video game, *Destiny.*

Find out more about Tee Morris at www.teemorris.com and at www.ministryof peculiaroccurrences.com.

Chuck Tomasi: In "real life," Chuck is currently employed as a Sr. Developer Advocate for ServiceNow, a software company providing a cloud-based workflow platform. Currently residing in Phoenix, Arizona, he is a devoted husband and proud father of two beautiful girls. As for his alter ego . . . Chuck is among the early podcast pioneers who started in 2004. He has produced over 1,000 episodes across multiple shows, the longest running being *Technorama,* a light-hearted geek show that he cohosts with Kreg Steppe. In 2011, Technorama took home a Parsec Award for best comedy/parody. He also cohosts *The Topic is Trek* with Kreg and podcast veteran Clinton Alvord. Chuck will tell you he's very fortunate to be able to take the skills and passion of podcasting and combine them with his technical acumen to deliver value to his employer. Since 2013, he has been the host and driving force behind *TechNow,* a ServiceNow developer web series, and *Break Point,* a ServiceNow developer podcast. He has also hosted/produced hundreds of developer-based audio and video productions. In 2020, Chuck hosted ServiceNow's CreatorCon conference, a weeklong digital event complete with keynote, workshops, breakout sessions, and a hackathon. When people talk about podcasting with passion, Chuck's enthusiasm comes to mind.

Learn more about Chuck at his website: www.chucktomasi.com.

Authors' Acknowledgments

Due to the complexity of the issue and the incredible growth in the community, it would be impossible to properly express our thanks to all the parties who were of great help with this book. So with that . . .

To our wives, Donna and Pip: Thanks for not strangling us for our constant "Oh! We've got to add that to the book!" and "Come on, honey, we have to record this for the book!" moments. We deeply appreciate the averted gazes of death when we answered that no, unfortunately we would not be coming to bed, no, I forgot to scoop the litter boxes because we had one more chapter to write, and that yes, we did realize it was 3 o'clock in the morning.

To the podcasters who provided not only inspiration, but also camaraderie and friendship along the way. Through listening to you all and talking to many, you served as a constant reminder of why we were pouring our hearts and souls into this text.

A huge thank you to Rob Walch, Vice President of Podcaster Relations at LibSyn and host of the podcasts The Feed and Podcast 411. Not only did you keep us in the loop concerning recent changes and updates at LibSyn, but you continue to be a constant supporter of the podcasting community. You are a friend and ally we can count on.

Finally, a special nod to Michael R. Mennenga for passing along an email back in 2004 that opened a door to a world of time-shifting, kick-ass mystic ninjas, podiobooks, and science-fiction and fantasy geeks around the world interested in what we have to offer.

Publisher's Acknowledgments

Executive Editor: Steven Hayes

Project Editor: Kelly Ewing

Technical Editor: Podcasting's Rich Sigfrit

Proofreader: Debbye Butler

Production Editor: Siddique Shaik

Cover Image: © Syda Productions/Shutterstock

Special Help: Donna Tomasi

Leverage the power

Dummies is the global leader in the reference category and one of the most trusted and highly regarded brands in the world. No longer just focused on books, customers now have access to the dummies content they need in the format they want. Together we'll craft a solution that engages your customers, stands out from the competition, and helps you meet your goals.

Advertising & Sponsorships

Connect with an engaged audience on a powerful multimedia site, and position your message alongside expert how-to content. Dummies.com is a one-stop shop for free, online information and know-how curated by a team of experts.

- Targeted ads
- Video
- Email Marketing
- Microsites
- Sweepstakes sponsorship

20 MILLION PAGE VIEWS
EVERY SINGLE MONTH

15 MILLION UNIQUE
VISITORS PER MONTH

43% OF ALL VISITORS ACCESS THE SITE
VIA THEIR MOBILE DEVICES

700,000 NEWSLETTER SUBSCRIPTIONS
TO THE INBOXES OF
300,000 UNIQUE INDIVIDUALS EVERY WEEK

# of dummies

Custom Publishing

Reach a global audience in any language by creating a solution that will differentiate you from competitors, amplify your message, and encourage customers to make a buying decision.

- Apps
- Books
- eBooks
- Video
- Audio
- Webinars

Brand Licensing & Content

Leverage the strength of the world's most popular reference brand to reach new audiences and channels of distribution.

For more information, visit dummies.com/biz

PERSONAL ENRICHMENT

Staying Sharp
9781119187790
USA $26.00
CAN $31.99
UK £19.99

Facebook
9781119179030
USA $21.99
CAN $25.99
UK £16.99

Guitar
9781119293354
USA $24.99
CAN $29.99
UK £17.99

Investing
9781119293347
USA $22.99
CAN $27.99
UK £16.99

Beekeeping
9781119310068
USA $22.99
CAN $27.99
UK £16.99

Digital Photography
9781119235606
USA $24.99
CAN $29.99
UK £17.99

Meditation
9781119251163
USA $24.99
CAN $29.99
UK £17.99

Pregnancy
9781119235491
USA $26.99
CAN $31.99
UK £19.99

Samsung Galaxy S7
9781119279952
USA $24.99
CAN $29.99
UK £17.99

iPhone
9781119283133
USA $24.99
CAN $29.99
UK £17.99

Crocheting
9781119287117
USA $24.99
CAN $29.99
UK £16.99

Nutrition
9781119130246
USA $22.99
CAN $27.99
UK £16.99

PROFESSIONAL DEVELOPMENT

Windows 10
9781119311041
USA $24.99
CAN $29.99
UK £17.99

AutoCAD
9781119255796
USA $39.99
CAN $47.99
UK £27.99

Excel 2016
9781119293439
USA $26.99
CAN $31.99
UK £19.99

QuickBooks 2017
9781119281467
USA $26.99
CAN $31.99
UK £19.99

macOS Sierra
9781119280651
USA $29.99
CAN $35.99
UK £21.99

LinkedIn
9781119251132
USA $24.99
CAN $29.99
UK £17.99

Windows 10
9781119310563
USA $34.00
CAN $41.99
UK £24.99

SharePoint 2016
9781119181705
USA $29.99
CAN $35.99
UK £21.99

Fundamental Analysis
9781119263593
USA $26.99
CAN $31.99
UK £19.99

Networking
9781119257769
USA $29.99
CAN $35.99
UK £21.99

Office 2016
9781119293477
USA $26.99
CAN $31.99
UK £19.99

Office 365
9781119265313
USA $24.99
CAN $29.99
UK £17.99

Salesforce.com
9781119239314
USA $29.99
CAN $35.99
UK £21.99

Coding
9781119293323
USA $29.99
CAN $35.99
UK £21.99

Learning Made Easy

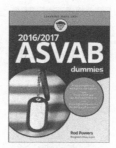

Small books for big imaginations

9781119177173
USA $9.99
CAN $9.99
UK £8.99

9781119177272
USA $9.99
CAN $9.99
UK £8.99

9781119177241
USA $9.99
CAN $9.99
UK £8.99

9781119177210
USA $9.99
CAN $9.99
UK £8.99

9781119262657
USA $9.99
CAN $9.99
UK £6.99

9781119291336
USA $9.99
CAN $9.99
UK £6.99

9781119233527
USA $9.99
CAN $9.99
UK £6.99

9781119291220
USA $9.99
CAN $9.99
UK £6.99

9781119177302
USA $9.99
CAN $9.99
UK £8.99

Unleash Their Creativity

dummies.com

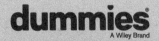